COOL WATER

By the same author:
Hot Water
Farewell Serena
Evil's Child

COOL WATER

Helen Rothman

D'Oro Press

This novel is entirely a work of fiction. The names, characters and incidents portrayed in it are the work of the author's imagination. Any resemblance to actual persons, living or dead, events or localities is entirely coincidental.

First published in Great Britain 2004 by
D'Oro Press
673 Finchley Road, London NW2 2JP

Copyright © Helen Rothman 2004

Helen Rothman asserts the moral right to
be identified as the author of this work

ISBN 0-9541552-3-8

Origination by
Pre-Press I.T., Weston-super-Mare, Somerset, Great Britain

Printed and bound in Denmark by
BookPartner, Copenhagen

All rights reserved. No part of this publication may be reproduced, stored in a retrieval system, or transmitted, in any form or by any means, electronic, mechanical, photocopying, recording or otherwise, without the prior permission of the publishers.

This book is sold subject to the condition that it shall not, by way of trade or otherwise, be lent, re-sold, hired out or otherwise circulated without the publisher's prior consent in any form or binding or cover other than that in which it is published and without a similar condition including this condition being imposed on the subsequent purchaser.

For my sister

Chapter One

The young man gazed down at the bay where his prey was located. He had chosen to come at dawn, and stand on the bluff overlooking the expanse of turquoise blue water. The tiny ripples shimmered in the rising sun, and he could make out a string of stingrays gliding along the sandy bottom, elegantly swaying their wing-like extremities. His eyes wandered over the fine, white, sandy beach, stretching inland towards the Casuarinas trees and the tropical vegetation. Twelve bungalows were dotted along the edge of the greenery. They were hardly visible, so well were they integrated into the landscape. Bach bungalow had a pool attached to it. The pools were irregular in size and looked like natural ponds, surrounded by multicoloured hibiscus bushes. There was an atmosphere of serenity which radiated about the place, a serenity which the young man had been ordered to destroy. He had been ordered to destroy Maurizio Capponcini and his family, owners of 'Serenity' the newest Club Royale venue, take the club over and name it Club Curria, to honour his dead half-brother, one Claudio Curria. Not only that, but also take over Petrolia, the tiny oil producing atoll, which had emerged from the ocean after the monster earthquake, five years ago, which had hit Tequila and the surrounding area. This then, was the vendetta which had brought him over to Tequila; the vendetta that his

Mafia family expected him to fulfil. Ostensibly he had accepted to fight a couple of bulls at the Semana Santa bullfight in Tequila on Palm Sunday. He was a fully fledged torero with a reputation, not only as a successful bullfighter, but as a bit of a womaniser, given his extraordinary good looks.

James Harvey, alias Caramba de la Cruz, picked up his binoculars which were hanging round his neck and scanned the shore-line. No yachts were permitted entry into Beatriz Bay. The only passe was very narrow, and there were only two boats in the bay. One was the club's speed boat and the other the club's sleek catamaran, anchored off shore, which was available for charter by the club's guests. He saw the man standing on the deck. James could just make out the smile of satisfaction on the man's face who surveyed the bay. A woman came to join him on the deck. Even at this distance he could see that she was a looker; tall, slim, with long shiny auburn hair falling over her shoulders. James whistled under his breath. Mr. and Mrs. High and Mighty Capponcini themselves, about to take out the cat for a sail, perhaps. They, and a few others, were responsible for his half-brother's ignominious demise. There was the girl known as the Chinese chick, her mother-in-law known as Old Crone Chang, the US consul, and last but not least, The President of Tequila herself, Maria Beatriz and that ridiculous Medicine Man commonly known as Sorciero, whom she had taken as her husband. James Harvey was a young man who didn't believe in the Island Voodoo. He was not one to be fooled by the local mumbo jumbo mouthed by Sorciero, or his legendary promenade on burning coals at the famous Happening which took place annually on Palm Saturday. James would participate at all the pre Holy

Week festivities, and although he despised all the hype that went with witchcraft, he was going to sneak a look at the Happening. Then he would strike. He didn't know quite how, but he would have to.

James wasn't an evil young man. He had been brought up on a ranch by a doting mother, a beautiful young American woman, who had married into the Curria clan. She had met James's father on a cruise and had married the man on board. The typical shipboard romance, which had very quickly lost its charm, when she discovered that she had plighted her troth to a serial bigamist. When she protested to her husband, she found herself bundled off and banished to an isolated ranch in Argentina with her unborn child. Vague, mysterious threats were uttered if she should ever leave the property. James was born on the ranch, was educated at home by his mother and a tutor, and became proficient in animal husbandry. The ranch produced fine fighting bulls, which were exported to the neighbouring Latin American states and the Caribbean. It wasn't surprising therefore, that James became a bull-fighter, took the name Caramba de la Cruz and travelled to the bullfighting arenas in Buenos Aires and Caracas. He had acquired quite a reputation, not only as a bull-fighter but also as a heartbreaker.

Over the years he had heard about his father's family, but never saw any of them. His father sent a 'Family' bodyguard to keep an eye on his progeny and keep the mother in line. Old Curria had an album of photographs portraying the boy's progress. He was secretly proud of this good looking young man, and glad that he had not been contaminated by the Family. There was no stigma attached to James, and who knew but that this might come in useful sometime for Family business. James's mother had always called him Baby, although he had been baptised James, Romeo, Giovanni, Batista, according to his absent father's wishes.

It was James's skill as a Torero which had lead to his being on Tequila, with a mission to accomplish. To destroy the owners of the Club Royale, and invade Petrolia; to avenge his half-brother's death, namely one Claudio Curia. In truth, James had no great desire to destroy anything except a good bull, but didn't know how to get out of this task. He couldn't have cared less about a father and a half-brother he had never seen; both unsavoury characters at that, according to his mother. Yet she had implored him to bow to his father's wish, and do his bidding. She was obviously terrified of the consequences if James failed in his mission. So here he was, Caramba de la Cruz, famous young Torero, who had been invited to participate in the legendary bullfight which took place on Palm Sunday on Tequila. His mother's bulls had been exported to Tequila for a number of years, so it was quite natural that James Harvey, alias Caramba de la Cruz should be welcomed with open arms. He would be the star of the Arena, except of course for the President herself who never missed a fight. Heavily pregnant with her first child, she had nonetheless participated in one of the previous Palm Sunday bullfights, dressed in a chic maternity bullfighters outfit, and having made her kill, had proceeded to give birth to a strapping male child in the Arena amid the jubilant cheers of her adoring audience. James hoped that she wasn't going to pull some other trick and rob him of his triumph.

He turned his attention to the catamaran again and watched another figure emerge on deck. He could see the young girl quite distinctly, her naked young breasts uplifted to greet the sunshine. Her auburn hair was a mass of ringlets streaked by the sun, and shone like a halo around her head. Her long shapely legs seemed to go on for ever. A tiny bikini bottom encased her firm small buttocks and a gold chain was tied around her waist. She also stretched her arms and smiled at the sky, although it seemed to James that she was looking straight up at him. He drew a sharp breath and felt a distinct twinge in his chest. It was immediately clear to him that he could no more harm this creature than fly to the moon. Indeed, it would be far more likely that he would fly to the moon. He didn't however immediately realize that he had quite simply and completely fallen in love with Miss Miranda Capponcini. All he knew was that he had to meet this vision as soon as possible, and with that in mind he removed the binoculars and stuffed them in his backpack, which he stowed behind a hibiscus bush. He nimbly slithered down the slope and slipped into the sea. He swam strongly out into the bay and then doubled back to the catamaran. As he came along-side he called out 'Ahoy there.'

Mauricio Capponcini bent over the side and looked questioningly at the swimmer. The swimmer lifted his arm and pointed towards the passe.

'Is that the way to Trinidad?'

Mauricio Capponcini smiled. 'Keep in this lane and then take the first turning right. After that you can consult the next shark that comes along.'

'Any chance of getting a little bit of sustenance before I go?'

'Are you staying at the club? I can't remember seeing you there.'

'My name is de la Cruz. I arrived from Argentina an hour ago. Any vacancies?'

'Bungalow five. Welcome aboard and have a cup of

coffee, de la Cruz.'

James climbed the ladder and landed on the deck. He was a fine looking young man, healthy in limb and well proportioned. He had the slender waist and the well turned ankle of a Torero. Only his colouring seemed odd for someone called Caramba de la Cruz. His fair skin and straight, silky, almost platinum blonde hair which fell to his shoulders, made him look more like a Viking than a Latin American bullfighter. His green, gold-flecked irises reflected the water and his lids were thickly fringed with darker lashes. Domitilla Capponcini came up from the cabin with a tray of steaming coffee and frothy milk. A basket of freshly baked mini Danish pastries was brought up by Miranda Capponcini, her eldest daughter. The girl had slipped on a skimpy T-shirt, and her oiled skin glistened like gilded bronze. James was hard put not to let his mouth drop open in admiration. He swallowed hard and gave a little bow to the ladies. Miranda smiled at the young God who had miraculously emerged from the sea. Her large hazel eyes locked into his green ones.

'Danish?' she breathed and held out the basket.

'No, Argentine,' he answered, still staring into her eyes. 'You have gold flecks in your left iris, just like I have. Has anyone told you that before?'

'It's nice to know that we have a gold fleck bond.' She smiled at him. 'However, people usually comment on my dimples.'

'That too. It's a completely unfair advantage to have both these extraordinary assets.'

Domitilla listened to this exchange with lifted eyebrows, and squinted over at her husband. He grimaced back at her and shrugged his shoulders.

'Coffee. De la Cruz?' he asked then.

'Thank you, Mr.Capponcini, but I've got some urgent business to attend to in Trinidad.' James held out his hand to Miranda and she took it. 'Come with me,' he said and stood at the edge of the deck. They looked at each other and started to laugh. It was a joyous sound. They jumped into the sea hand in hand, and came up for breath still laughing.

'Don't forget that you're taking the yoga class at 9.30, Randy,' Domitilla called after her daughter. The pair swam away with rhythmic strokes, both excellent swimmers.

'Well my love, we'll soon have to console another discarded suitor,' Maurizio commented.

'I somehow don't think so. This is the legendary love at first sight syndrome.'

'Do me a favour; what would Randy want with a bullfighter. Besides what would you know about that syndrome?' He put his arm around his wife.

'I'm speaking from experience. Happened to us, remember?'

'How could I forget; but that was us, not Miranda and some outlandish Argentine Torero.'

'I wouldn't knock him if I were you,' Domitilla smiled quietly. 'Let's have our breakfast before it gets cold.'

James and Miranda reached an outcrop of rocks where a discreet little plateau had been built for solitude seeking sunbathers. He scrambled up and held out his hand to her. She clutched it and hauled herself out.

'How did you know about this place?'

'I saw it from the bluff, over there.'

'So you were watching us?' He looked embarrassed and didn't answer. 'You saw me on deck?' He nodded. 'With binoculars?' Again he nodded.

'Then you won't mind too much if I take off my T-shirt, I hate wet clothes on my body.' She stripped off the wet shirt and the bikini bottoms, and lay down naked. 'The sun feels good in Trinidad,' she murmured and closed her eyes.

'You're the most beautiful girl I've ever seen,' he blurted out.

'And you're the most beautiful man I've ever seen,' she murmured. James looked around in desperation. He felt his penis grow and bulge out impatiently, urging him to action.

'I've got to go,' he said and dived back into the water. She sat up and gave a small wave.

'Cold water always does the trick, especially with dogs, you know,' she said earnestly. 'Coming to the Yoga?'

'No, but I'll take you out to dinner. Nine o'clock OK?' He got hold of her foot which was dangling in the water and kissed her toes. He turned and swam towards the bluff with turmoil in his head and excitement in his heart.

The backpack was still hidden among the bushes, and he picked it up and took the path down from the bluff into the gardens of the Club Royale. He thought it might be better to use the main entrance rather than swim in from the beach to claim Bungalow 5. An attractive young Tequilan girl was at the reception desk in the lobby. An arrangement of fresh orchids decorated the desk and cascaded down the front, forming garlands of immaculate white blooms. James stood in front of her and she raised her perfectly plucked eyebrows.

'I would like to have the key to Bungalow 5, please,' Baby said.

'You must be Caramba,' she breathed, her eyes as big as saucers. 'I've seen your picture in "OLA"'.

'So you know about me?'

'Mr Capponcini phoned in to tell us to expect you.'

She batted her eyelids hard. 'I would have recognized you anyway.' She pushed a button. 'The bellboy will show you to your bungalow and take up your luggage.'

'I haven't got any, except for my backpack. It's gone astray.'

'Oh dear, what are you going to do?' The receptionist pursed her full lips thoughtfully; then she had a definite brainwave which brought the smile back to her lips.

'You'll find anything you need in our club shop, from a bathing-suit to a dinner-jacket. We don't really have much call for such formal attire on our island, except of course for the Government Ball on Palm Sunday, you know, and the masked ball tonight, here at the club.'

'Does the club shop carry a complete torero outfit for my appearance at the bullfight?'

The receptionist gasped her distress. 'Oh, Senor de la Cruz, I don't think so, but my mother is a wonderful seamstress and could run it up for you.'

'You really are a most helpful girl, but no thank you. I expect it will arrive in time for the bullfight. Any chance of getting some breakfast?'

'Oh yes, of course. Just order it from room service, or take it on the terrace. They are starting to serve now.'

He looked down at himself and shook his head. He took another look name tag. 'I think it's got to be room service, Jolie.' He smiled at her. 'Nice name, suits you.'

'Oh thank you, Sir,' the girl flushed with pleasure. 'There's Juanito, he'll show you to No.5.'

The bungalow was spacious and ingeniously furnished to give the maximum space and comfort. The colour schemes were fresh and soothing. There was a bottle of iced mineral water, crystal goblets, and a dish of tropical fruit on the coffee table. The refrigerator, which was concealed in an antique oriental cabinet, was stocked with champagne, wines, spirits and water. The adjacent bathroom was like a green house, planted with flowering bushes and miniature palm trees. It was fitted with an electrically operated glass roof which left it open to the sky. A classical terracotta sculpture of a woman, stood among the greenery carrying a basket with all the usual complimentary soaps, perfumes, lotions and gels. James dropped his backpack and his clothes and activated the shower. Hot water gushed generously from the gleaming shower head and he stepped under it, sponging the salt water off his body. It felt good and he hummed contentedly while he soaped himself. The bathrobe hung by the shower door, and he put it on. It felt soft and fluffy and smelled of fresh lavender. He picked up the phone and put his order through for breakfast. Then he threw himself on the king size bed and gazed at the ceiling. His feeling of contentment would have been complete, had it not been for the niggling thought which reminded him of the true purpose of his stay on this blessed island. It intruded, unbidden, from his subconscious. He groaned as the thought took over completely, and he lay there frozen with horror at the prospect. He barely heard the discreet knock on the door, and a soft voice telling him that his breakfast was laid on the porch. James picked himself up and looked for his backpack. He undid the fastening and groped around in it. Apart from his tickets, passport, money and basic toiletries, it contained a large, hard bound copy of the Bible. He took out the Bible and activated the secret spring he had been shown before his departure. The book had been holed out and he stared at the hand gun which had miraculously found its way into the Bible. He looked at it with loathing but picked it up and held it. The door to the bungalow flew open, and Miranda Capponcini let out a little scream.

'Why in heavens name are you pointing a gun at me?' she demanded.

'I... I... am?' he stammered and shoved it in his pocket.

'Or is it just a cigarette lighter, like in the movies.'

'I don't smoke,' he said.

'Then it's really a gun?' she sounded awed.

'Yes,' he admitted reluctantly. 'Rumour has it that there is quite a lot of crime on the island, and it was suggested I carry it for my own protection.'

She giggled and shook her halo of hair. 'What a lot of nonsense; the island's as safe as houses since the Mafia was evicted and Maria-Beatriz is President. Things have radically changed here; the discovery of oil on Petrolia has brought prosperity, you know? The earthquake destroyed the shanty towns and now every family has a reasonable home and the new hospital is really taking care of the sick. The main roads around the island have been asphalted, there's no unemployment, the new schools are modern and there are good teachers from the USA and Spain. Education has really improved, and the University will be open soon.'

'Sounds like Utopia... almost too good to be true,' James said skeptically.

'It's true alright; you'll see when I take you around the island.'

'And when will that be?'

'After my yoga and your breakfast. Are you game?'

'I have to wait until the shop opens and buy some clothes. No luggage, you see? It apparently never came on the same plane.'

'Shall I send something around for you to wear? I've got the key of the boutique.' She looked at him questioningly, and he nodded his agreement.

'By the way, I can't call you Caramba, Caramba. That's not really your name, is it?' He looked down somewhat shamefacedly and shook his head.

'My name is James Romeo Giovanni Batista Harvey,' he said.

'It's a nice selection of names, so why the embarrassment?'

'I have a nickname... That is, my mother always calls me that.'

'Confess,' she commanded.

'Baby,' he said reluctantly.

'Isn't our acquaintance rather recent for you to call me that?' she demurred.

'Not you, I'm Baby,'

Her laughter burst out and she touched his cheek. 'Oh, Baby, tell me it isn't true.'

Then he was laughing as well and he kissed her hand. 'Oh, Baby, it is. but generally I am known as James.'

'I think I'll call you James. Would you like that?'

'I'm sure that I will like anything you say or do; what do I call you?'

'Randy. See you in about an hour's time.'

He held open the door for her and watched her briskly jog up the path to the main building.

He sat down to his breakfast. It was not only a feast for the eyes, but also a feast for the animal which growled in his stomach. The scrambled eggs were fluffy and soft, heaped onto a crisp piece of toast, topped by a cherry tomato. A dish of beautifully cut tropical fruit arranged in a fan pattern, a basket of fresh rolls and brioches wrapped in a starched white linen napkin, a jar of honey and a jar of homemade strawberry jam were placed on the white and blue striped starched tablecloth. James poured out the coffee from the thermos. The aroma was pungent and sweet at the same time. He inhaled it with pleasure. Life was good, he told himself, better than he had ever expected. With the typical insouciance of youth, he was sure that he would somehow manage to get himself out of the quandary he was in.

She drove well, keeping an even tempo. He sat beside her wearing his new pale blue linen shorts and matching shirt with the tiny logo of the Club Royale embroidered on the sleeve. His fine long blonde hair whipped around his face and got into his eyes. She had tied a bandanna over her head, keeping her halo in check. They rounded a bend and she stopped in a parking area. The view over a large bay was breathtaking. They got out of the Land Rover and stood at the edge of the cliff. A flight of frigate birds cruised above them and drifted out over the sea.

'It's almost paradise, James.'

'Except for the oil rigs over there.'

'It does spoil the view a little, but think of the good they do,' Randy said earnestly.

He looked into the distance and could make out the islet which was surrounded by the rigs. 'So that little itsy bitsy dot in the vast Ocean is Tequila's money spinner,' he murmured.

She nodded. 'That's Petrolia.'

'Is it within Tequila's territorial waters?' he asked.

'Strictly speaking, it's a borderline case, but Tequila laid claim to it. and no other island, or mainland for that matter, is closer.'

'So by law, it is Tequilan territory?'

'Possession is nine tenths of the law, and Tequila is firmly entrenched on it.'

'I'd love to go and take a look at it,' he said.

'A special permit is required, and Petrolia is heavily

guarded. The government is still afraid that the Mafia might want to take it over, after what happened here just before and after the earthquake.'

'What happened?' he asked idly.

'The whole Mafia drug and gambling operation was blown sky high on the island. It is rumoured that it was not only the earthquake which blasted them off the island, but that the CIA played a part in the destruction. A prominent member of the Mafia perished with a number of his minions on the island, plus their super yacht which they used for transporting the drugs they manufactured here. The man's name was Curria and he was found dead on Petrolia by my father and Mei-ling Chang, who runs the local dive shop.'

James swallowed hard and then found his voice. 'Did they kill him?' he wanted to know.

'Course not. My father couldn't harm a fly. They were out on a night dive, and were thrown onto Petrolia, as it emerged from the sea during the earthquake'

'Quite a story,' James murmured. 'I've heard a rumour that this Mei-ling had a hand in all this. I've heard that she was a CIA agent.'

'Ridiculous; she looks like a lovely Chinese doll, small and dainty, not like a hit man. Let's get on with our tour of the island. Want to drive?'

He shook his head. 'No way, I enjoy being driven by a good driver.'

'That's a fine compliment coming from someone who usually drives a Lamborghini like a bat out of hell.'

'How on earth do you know that?'

'Jolie insisted I look at the OLA. She gets it every week and has kept all the copies for almost two years. It's how she keeps track of some of our guests, you know; the rich and infamous. It's quite helpful sometimes knowing a future guest's likes and dislikes. You were in it some weeks ago, with your arm around a beautiful bimbo, burning up the dirt track on your estancia.' She squinted at him and he grinned, ignoring the dig.

'How old are you, Randy?'

'Has no one ever told you that you never ask a lady her age or weight?' she rebuked him.

'Let me guess; twenty, and 130lbs. Vital statistics. Let me see; thirty-six, twenty-two, thirty-four. Height about 5ft.7ins.'

'How about my blood group?'

'Hot, I should say.'

She drove on the main highway which was bordered on both sides by acacia trees. Neat little cottages and bigger houses were visible in the distance. Papaya trees with plump fruit and banana plantations spread over the hills. Fields of the tequila producing agaves sprouted all over. The sun was hot now and he squinted at the sky. She wordlessly gave him a pair of sunglasses which she retrieved from the side pocket. As the road snaked up the hill, it ran through thickets of rain forest. A waterfall splashed down the rock face, filling a pool of clear water. Water lilies grew on the edge, and strange giant leaves, waxy and shiny, stood guard on the banks of the stream which flowed out of the pool down to the valley. It was steamy and hot under the cover of the enormous old trees. She stopped at a bridge which crossed the stream. It was built out of mahogany, intricately carved, and dainty.

'This is almost my favourite place. I find it hard to decide which spot like best here. Do you know that this bridge is supposed to be at least a hundred years old?'

He looked at her face which was filled with joy at the beauty of their surroundings. 'You sound as if you would like to spend the rest of your days here,' he said.

'I would like to, but it's not reasonable. I have to finish my studies at the university in Milan.' She smiled at him ruefully. 'In fact, my holidays are almost over. I leave immediately after Easter.'

'But you can't do that,' he protested.

'Aren't you leaving as well?'

I'm going to stay at least another week after Easter, for a short vacation. I could stay longer...'

'Let's drive on, there's still a lot to see,' she cut him short and returned to the Land Rover. They drove on and soon were at the end of the rainforest. They drove past a large new building. Nurses in white were walking in the gardens with patients.

'The hospital,' she mouthed. He pointed to a group of buildings further down.

'The elementary school and secondary school. The high school is a little further away.' He nodded and admired the neatness of the playing fields, and the small sports stadium. A group of youngsters were playing soccer, while others were running in the stadium. She slowed down and drove through the small city to the waterfront. A colossal cruise liner was moored at the dock. Young bicycle rickshaw drivers were vying for fares. These rickshaws were still used in the inner city, but the public transport service had been greatly improved. Compact new buses reached every village on the island, and kept to a strict schedule. Hotels were still low rise and comfortable, built so as not to interfere with the landscape.

'Where's the airport from here,' he asked. 'I could check whether my luggage has miraculously arrived.'

'It's about five miles across to the other side. Would you like to have lunch at the Yacht Club? They have quite a decent buffet lunch.'

'Sounds good to me,' he agreed. She turned the Land Rover around and soon they were entering the airport area. She drew up in front of the large hall and he got out. She pointed to a side entrance. 'Ask over there,' she advised him. 'I'll stay in the car.' He got to the door and walked into the air-conditioned office. The girl looked up from her computer and smiled.

'I've come to inquire about my luggage.'

'What name, Sir?'

'Caramba de la Cruz. I arrived on the Inter-island Airways, this morning at 5.30 from Venezuela.'

'This must be your lucky day. Senor de la Cruz. We found it. The luggage was accidentally deposited in the transit area. A passenger who arrived from Miami a little while ago, offered to take your luggage to the Carlton Hotel. I believe you're staying there?'

'Not any more. I've decided to book into the Club Royale instead.'

'Oh, very nice indeed,' the girl was obviously impressed. 'Well, you must go and pick your luggage up at the Carlton. Your friend, who took the luggage, is also staying there.'

James' eyebrows shot up. 'A friend of mine? What's his name?'

'He definitely said he was a friend of yours,' she said and scanned her computer. 'That would be a Mr. Medusa. Muso Medusa, he said.'

'How odd, never heard of the fellow.'

Randy drove him to the Carlton Hotel. It was on the other side of the island. It had been the premier establishment before the creation of the Club Royale. It was an old hotel which was beautifully situated on a calm bay, having another, rougher water on its other side. It was just beginning to show its age and was in need of refurbishing, but had great charm. It was set in an old wooded park and had its own stream, which meandered into the bay. There a was nine hole golf course as well as some tennis courts. It was an ideal place for families with children and had all the water sports available, as well as a facility for childcare. Randy drove though the gates, down the tree-lined gravel road, and drew up in front of the hotel.

'Are you getting out?' Randy asked. James looked at the entrance thoughtfully. A tall, corpulent gentleman, sporting colourful tartan shorts, a bright red linen blazer and a Panama hat, stood beside two large Louis Vuitton cases. He was chatting to a uniformed bellboy. The bellboy came to open the car door, but James waved him away.

'There's something strange going on here, Randy. Some guy, who says he's a friend of mine, picks up my luggage and brings it here. His name is supposedly Muso Medusa. I've never heard of him.'

'I'm glad you haven't. It sounds exactly like a Mafioso name.' James beckoned to the bellboy who advanced again. 'Could you bring me the two cases standing by the entrance? They're mine.

'I can't do that, Sir; you had better identify yourself at reception first.'

'Just tell reception that Caramba de la Cruz is outside and would like to have his luggage.' The bellboy shrugged his shoulders and went into the lobby. The gentleman at the entrance lifted his Panama hat and fanned himself with it. He watched the two young people in the Land Rover closely. Then he decided to lift the weight off his feet and sat down on one of the cases. James shot out of the car and marched up to him.

'Be so kind as to lift yourself off my luggage,' James said sharply. The man jumped up, and his face creased into smiles. He wrapped his large arms around James and hugged him to his ample frame.

'Caramba, am I glad to see you. I thought that you had been kidnapped. You were gone, and your suitcases were standing there, abandoned at the airport.'

James tried to struggle out of this unwelcome embrace. 'Who the hell are you?' he asked, hardly bothering to mask his distaste.

'I'm Muso Medusa, you know, a friend of your father's. My, you've sure changed since I last saw you, down on the ranch.'

'I've never seen you in this life or any other. And you certainly were never down on the ranch.'

'You were a toddler then, so I won't blame you for not knowing me.' Muso Medusa smiled, showing an array of gleaming gold crowns. 'Come in, you've a nice suite here, I've inspected it for you.'

'Mister, whoever you are, what right have you got to inspect my suite? Out of my way.' James finally managed to push the fat man to one side and went into the lobby. He got out his passport at the reception and explained that he had come for his luggage. The receptionist was nonplussed.

'Your suite is ready for you, Senor de la Cruz. Your friend said that you would like it.'

'What I want is my luggage. Mr.Medusa is welcome to the suite if he likes it that much'.

'But, Sir, your reservation, the account?'

'Just give to Mr.Medusa.'

The bellboy loaded the two cases into the Land Rover and accepted the tip that James pressed into his hand.

'Where will you be staying, Senor de la Cruz, in case there are any messages or mail?'

'I'm not expecting anything. If it does, I'm certain Mr.Medusa will be happy to take care of that as well.'

'Any chance of getting a ticket for the bullfight? The staff here were hoping that perhaps...' the bellboy winked significantly.

'Sorry, I have nothing to do with the tickets, but what's your name?'

'Rodrigo, Senor.'

'I'll see what I can do.'
'Where will you be staying?'
'Nosy, aren't you? Let's drive on, Randy.'
'OK., boss. Ready to go to lunch? They'll stop serving in twenty minutes.'

She drove through the town again, manoeuvring skilfully down the narrow streets which led to the harbour, and stopped at the Yacht Club. She parked the Land Rover and they stepped onto the terrace where the tables were laid for lunch. It was pretty crowded with regular clients, and the usual yacht crews holding up the bar. Miranda waved to a young Chinese couple who were sitting at a large table. They waved back and motioned for Miranda to join them.

'Mei-ling and Young Chang have asked us to sit with them. We might as well, the place is packed. They're very nice and good friends.'

'Is she the diving Diva of the island?'

'Yes, she's the one who was washed up on Petrolia with my Dad. Let's get something to eat, before they take all the goodies away.'

'What are we planning to do after lunch?'

'You're indefatigable, it seems,' she sighed. 'You'll find that a long siesta will be very beneficial. My folks are throwing their annual masked ball tonight. Anybody who's anybody will be there, even the President herself with her husband. So come to reception at 7.30 p.m. No shorts or T-shirts.'

'Yes, Ma'm, and then can we have dinner just you and I, by candlelight, with soft music, assuming there is such a place.'

'There certainly is such a place. I'll take care of the booking.'

'How is it possible that a girl like you didn't have a dinner engagement for tonight?'

'Oh I did, but I cancelled it after breakfast.' She smiled at him, and he felt his heart turn a somersault in his chest at the sight of her dimples.

'Thank you,' he said simply and kissed the tip of her nose.

'You can't do that here, in public,' she protested. 'Only if you are engaged to be married, that is.'

'Will you marry me, so that I can do it again? I find it irresistible.'

'Of course I will, after we've had lunch. I'm starved.'

They helped themselves to the seafood and salads, and then joined Mei-ling and Young Chang. Miranda introduced James and soon they were all in animated conversation about the events which were planned for the Semana Santa. They both promised to come to the corrida, to see James do his bit in the arena; and of course, they would all go to the big Happening in the clearing in front of Sorciero's old house, which had been miraculously left untouched by the earthquake. James smiled but his unease grew. Now that he had met the Chinese chick, another one of the conspirators he had been ordered to destroy, he immediately knew that he could never harm her either. He still had to meet the American Consul, Jason Reed, and Young Chang's mother, the almost legendary herbalist, Old Crone Chang. Last but not least on the list were the President, Maria Beatriz, and her witchdoctor husband, Sorciero. A queasiness suddenly overcame James at the thought of the prospect, and he sprang up and asked to be excused. He almost ran to the men's rest room and sat down, holding his head in his hands. He didn't often pray, but now he fervently prayed to God to send him the solution to his problem.

Chapter Two

Old Crone Chang and Sorciero were sitting inside his old house in companionable silence, as they always did after their consulting sessions. The old Chinese matriarch and the local witchdoctor Sorciero were as unlikely partners as one could imagine, but they had started their practice together a number of years ago, and dispensed herbal remedies, magic potions and spells. Maria Pilar, their assistant, had gone home for the day. A high speed fan was keeping the afternoon heat at bay and they were sipping Old Crone's special brew. A herbal tea composed of secret ingredients, which fortunately no one would have dared to analyse. Old Crone had insisted that her little patch of herb garden was out of bounds to everyone except herself and Sorciero. The tender plants of Marijuana which grew there, undisturbed, were a considerable part of those ingredients, and the tea was otherwise dispensed only to the terminally ill. When they had finished the tea and put down the delicate Chinese porcelain cups, Sorciero brought out the cash box and proceeded to count the day's takings. There were as usual, a number of one dollar bills, with the occasional tens. A cage standing in the corner of the room contained a couple of laying liens. A bucket with some good size snappers was next to it, as well as a finely woven basket which was filled with small pineapples. A chocolate gateau, garishly iced in pink, took pride of place on the table. Sorciero looked at their earnings and shook his head.

'I don't know, Old Crone, I don't know whether it's worth keeping on with this business. There won't be much left over after we've paid Maria Pilar.'

'What you say is true, Sorciero, but I wouldn't give up your profession just like that.'

'Maria-Beatriz wants me to give it up and help her more with her commitments, and hopefully there soon might be another baby on the way; when that happens, she'll need me more.'

Old Crone seemed to give this some thought before answering.

'Listen to me, Sorciero, don't do it. Help her on the social side, by all means, but she'll probably need more help in the nursery. Ling-Ling has a sister who might come over from San Francisco to help in the nursery when the time comes.'

'That's a good thought, but I still think this business is not worth our time, Old Crone; the takings don't warrant it.'

'Sorciero, this has nothing to do with money, although I don't turn my nose up at a good laying hen, or some freshly caught fish. What is important here, is that we get all the news. We hear all the rumours, all the scandals, in short everything that is happening on the island. It gives you power, Sorciero; don't let that power get away, it's an incredible advantage.'

It was Sorciero's turn to mull over Old Crone's opinion.

'You're a cool, wise old bird, Old Crone. I'll discuss it with Maria Beatriz, she'll understand.'

Old Crone nodded her head in approval. 'You are a bright young man, Sorciero and a true witchdoctor. You have the gift, don't squander it.' She rose from the high-backed mandarin throne, which had been her contribution towards furnishing the practice, and smoothed her black silk tunic. She patted down her jet black hair which was just beginning to be threaded with silver.

'Do you want to share the hens and the fish?' he asked.

'With pleasure, if you can help me put it in my rickshaw. You take the pineapples; I know that the President loves them. By the way, you know what Rodrigo, the bell-boy from the Carlton told us about the gaudy American with the Italian name. I wonder if you should tell the President about him. He may be bad news.'

'Will do, Old Crone.' They both turned towards the screened door, when they heard a vehicle crunch to a stop on the gravel path. It was too late for consultations. Everyone on the island knew that Sorciero and Old Crone shut shop at four p.m.

'Go and have a look, Sorciero, please, and tell them to come back tomorrow,' Old Crone Chang murmured. Sorciero opened the screened door and stepped out on to the porch. He looked in wonderment at the white stretch limousine and the motorcycle police escort which had drawn up outside the modest house, which had been his home before he had married the President. A flag on the limo designated it as an official vehicle, but it wasn't the Tequilan flag. Sorciero immediately recognized it as the flag of a neighbouring island, which on a clear day could be seen to the South West. It was a delightfully pretty island, with exquisite beaches, but through some quirk of climatic conditions, it lacked water. The rainfall seemed to stop just a few yards off shore and when the rest of the Caribbean was drenched, Mangorenia remained sunny and dry. Its only money spinner was the yearly crop of a tropical fruit which had been grafted and cultivated there, named Mangoreen. It was watered with water reclaimed from the sea by a small desalination plant. It was not enough however, to sustain a growing population, and the correct infrastructure for tourism remained a project for the distant future. Mangorenia stayed desperately poor and backward, although it did have some of the outward trappings of prosperity, like the Presidential limo, which had pulled up outside Sorciero's house. It also was represented at the United Nations, just like Tequila.

A handsome middle-aged man stepped out of the limo. His impeccable linen suit hung well on his tall frame and the Hermes tie had been carefully selected to match his shirt. He embraced Sorciero repeatedly.

'My dear Sam, how nice to see you again,' the man said heartily. 'You remember the last time we met at the UN in New York? I trust you're coming to the plenary session in the autumn. We need to remind them of the plight, the abject poverty of the Caribbean.'

Sorciero repeatedly embraced the President of Mangorenia. 'Yea, yea, Mun, we need to remind the whole world.'

'As well as our nearest neighbours,' the President said earnestly.

'Etienne, is this a hint, Mun? We've done our bit; the canning factory for the Mangoreens, and the Marina, remember?' Sorciero sounded a trifle aggrieved. He felt somewhat at a disadvantage however, confronting this elegant visitor. Sorciero was still dressed in his working clothes; traditional body paint covered his naked chest, and a skimpy loincloth concealed his manhood, but it was not exactly the right garb in which to receive a President.

'How could I forget, my dear boy? But we need more help.' Sorciero was about to reply, but the President held up is hand. 'Can I come in, so that we can talk?'

Sorciero nodded and led the visitor through the screened door. The President bowed to Old Crone Chang. She looked even smaller and frailer than usual standing next to the President.

'May I introduce Madame Chang, Etienne? We work together, dispensing herbal potions and remedies.'

'I have heard a lot about you, Madame. Delighted to meet you.' Old Crone inclined her head and took her usual little steps to the door.

'I am honoured, Mr. President. I must go now, Sorciero, if you'll excuse me. I'm sure you two have a lot to talk about.'

President Etienne Bonheur took the seat which Sorciero had quickly dusted off and offered to his unexpected guest. He wondered what Etienne Bonheur would request this time. Ever since Tequila had struck oil, all the neighbouring islands were sending emissaries, with a view of picking up some handouts. Sorciero donned a colourful robe and removed his feathered headdress.

'Etienne, do you want something to drink?'

Etienne Bonheur smiled and crossed his knees. 'I've heard that you and Mrs. Chang dispense a good cup of tea.' Sorciero nodded and put the kettle on to boil again. He produced the fine bone china cups used for the tea. He measured out two spoons of the tea into the teapot and waited for the President to speak.

'We need to find something to enhance our economy, Sam.'

'Bad harvest of Mangoreens?' Sorciero asked.

'Mangoreens are doing as well as can be expected, but we need something bigger and better, Sam. Ever since the earthquake we have had a population explosion, and we can't figure out why.' The President looked at Sorciero seriously and crossed his elegant knees.

'That's not too difficult to figure out, I suppose you guys just like fucking,' Sorciero said.

'It's alright for you to scoff, Sammie, but I'm serious. We are going to find it hard to feed our people, unless we find oil, or something. I've heard that you are a diviner, that you can find things buried in the earth,' he continued.

'I have discovered a few things in my time. I found a gold cache which had been buried for centuries by some pirates.'

'I heard about that... that's why I'm here.'

'Do you have some information about a possible gold cache on Mangorenia?' Sorciero asked, his eyes now bright with anticipation.

The President pursed his lips. 'We have found a document about a Spanish galleon which is supposed to have gone aground in our Lonely bay, but that's not the purpose of my visit.' He stopped and looked at Sorciero who had picked up the kettle. "We need oil, Sam.'

'Well, you can't have ours, so forget it.' Sorciero poured the water into the teapot and sat back on the Mandarin throne.

'We are not attempting to invade Petrolia, Sam, so relax.' Sorciero grunted and poured the tea. He handed Etienne Bonheur a cup and they both sipped the hot liquid. Sweat broke out on the visitor's brow and he wiped it with the finest of fine linen handkerchiefs. Sorciero turned the fan to maximum and waited for his guest to make the first move.

'Strange taste, your tea,' he said.

'Don't worry, it won't poison you. So you want to talk about oil.'

'I want you to look for oil. It seems to me and my government that there could be some found on Mangorenia, it being quite close to Petrolia.' Sorciero looked into his cup thoughtfully not knowing quite how to answer. He didn't know what his wife, the President of Tequila would think of the idea.

'We will pay you handsomely, of course.' Etienne Bonheur said. Still Sorciero didn't commit himself. 'Even if you're not successful in the quest,' Etienne Bonheur went on.

'How much,' Sorciero asked baldy.

'Five hundred dollars a day, for one week.'

'I can't guarantee a strike in one week. Mangorenia is small, but not that small. There's a lot of territory to cover.'

'We're willing to take a chance, so is it a deal?'

'When do you want me to start?' Sorciero asked.

'As soon as possible. Tomorrow, if you can make it. You could hitch a ride with us.' Sorciero put down his cup and scratched his head.

'More tea, Etienne?' he asked playing for time.

'Powerful stuff, Sam. It's made me feel good. Can I take some home?' Sorciero shook his head regretfully.

'I can't bend the rules, not even for you, Etienne. About this divining business, I'll have to consult Maria Beatriz, you understand?'

'Sure, sure, but don't take too long.'

'Why don't you get some professional geologists, Etienne? Venezuelans are supposed to be very good.'

'They've been, Sam. Been there, done that, without success. You are our only hope now. How about you let me know tonight at the Club Royale party. I take it that you and your lovely wife are both going there tonight.'

'Sure thing, Mun. Nothing like the Club Royale party to start the pre Semana Santa celebrations; the Capponcinis sure know how to throw a rave. It's a masked ball this time.'

Etienne Bonheur put down his empty cup and got up. The two men embraced again, and Sorciero showed his distinguished visitor out. The outriders got back onto their bikes and the white limo crunched over the gravel path to the main road. Sorciero gave a last wave, and then hurried back into the house. He discarded the robe and changed into his shorts and T-shirt. He was anxious to get home and discuss the proposition with his wife. Not that he needed the money so badly. His share of the practice gave him enough pocket-money to be getting on with but to earn several thousand dollars in a matter of days was an attractive prospect, because although he was the husband of the President, he held no official position as a civil servant, and had no other form of income. Gone were the days of graft, corruption and kickbacks for high ranking government officials, their families and friends. He turned off the fan and packed his feathered head-dress carefully into a plastic bag as well as his pouch of body paints. He was going as himself to the masked ball; Sorciero, the Witchdoctor of Tequila; a man, if not of substance, then of true sorcery, respected by all the islanders.

Old Crone Chang dismissed her rickshaw boy in front of her bungalow, and walked through the front garden. Her favourite flowering bushes were growing there in great profusion. Bougainvillea, Hibiscus, Oleander and Frangipani. The scent of Frangipani hung in the still air of the late afternoon. She unlocked the front door and gave a little sigh of contentment. She was always overcome by a warm feeling of pleasure when she entered her domain. She had worked hard and schemed tirelessly to achieve her independence. Her eyes drifted lovingly over the antique Chinese lacquer furniture, the bureau bookcase filled with early Chinese porcelain and pottery, the authentic Tang horse standing on the chest of drawers, and the Jade. Two intricately carved birds took pride of place on the dining-room table, between them, a luxuriant spray of white orchids in a famille noire jardiniere. Old Crone smiled with delight. Her daughter-in-law Mei-Ling must have been in the house and brought the orchids. Old Crone was very fond of her daughter-in-law, and it had given her great satisfaction when Young Chang had fallen in love with the girl and brought her home as his bride. The only sadness was that there still was no sign of the grandchild she would so dearly liked to have had. Mei-Ling told her to be patient, but the old lady found it hard to wait. She would have to ask Sorciero for a bit of his magic, if nothing happened soon. She went to the kitchen and looked out at the back garden. It was here that she grew the herbs necessary for her potions. No one was allowed to look after those. Not even Mei-Ling. Old Crone glanced at her watch. It was an antique jewelled and enamelled fob-watch which hung on a gold bow pinned to her tunic. It was time for the sprinklers to come on, and she waited until they did. Only then did she go and sit down on her favourite chair and closed her eyes for a short rest, before getting ready for the first of many parties which were given during the following week.

Those seven days before Palm Sunday were traditionally celebrated in a very Tequilan way. Masses were held at the Cathedral, the annual bullfighting festival took place, and Sorciero's famous Happening pulled in aficionados from all over the Caribbean. The Government Ball given every Palm Sunday night at the Presidential pink marble Villa was the climax. Small jets landed constantly during that week, bringing in celebrants. The marina was jammed packed with ocean-going yachts, and the Yacht Club gave the Friday dinner which was attended by the Harbourmaster and was a freebee for all those who stayed on the yachts. It was an exhausting timetable and at the end of the celebrations the island became eerily quiet. The small jets had flown off and the yachts were leaving the Marina and sailing out to sea. The streets were deserted, the market stalls empty. Monday, in the late afternoon the island would stir again, and the Spanish style paseo down the main street to the harbour and back would start again. Some food stands would open and would start their barbeques. There would be a strong smell of garlic sausage and grilled fish lingering over the streets. Life would be getting back to normal. Tequilans were deeply religious and Holy Week was a time for going to the Cathedral for prayers and mourning for Jesus, and joyous celebrations for his resurrection.

Mei-ling and Young Chang put on their traditional Chinese garments. Young Chang wore an antique Mandarin Coat, richly embroidered with gold thread and many coloured silks. Mei-ling wore a pale lilac chungsam, which fitted her small well shaped body perfectly. Herons were woven into the jacquard silk, in various shades of lilac. The cloth had been lying in Old Crone's marriage chest for more years than she cared to remember. She had lovingly folded and refolded it many times during those years, and layered it with tissue paper, so that the silk would not rot in the creases. She had sown up the chungsam herself with laborious tiny stitches as a gift for her daughter-in-law.

Young Chang looked at his dainty, beautiful young wife as she swept her hair into a shiny chignon. He sighed with pleasure, and thought, as he did every day, how lucky she had been to survive the earthquake, how lucky he had been that she had fallen in love with him. He went to stand behind her and placed a jewel box on the dressing-table.

'Not another one, Darling,' she chided him gently.

'I just couldn't resist, my love,' he said. She opened

the box and smiled with pleasure at the contents. A pair of antique, three-pronged tortoiseshell hairpins, encrusted with diamonds and amethysts lay on the velvet cushion. She picked them up and gave them to him.

'They're perfect, Darling. Would you put them into my hair?' He slipped them in her jet black hair, and stood back to admire the effect in the mirror.

'How can I ever thank you,' she murmured.

'By keeping your promise about giving up your job. You know how I worry about you. Every time you go diving, I fear that you'll not come back to the surface.'

She got up and put her arms around him. 'I love diving; it's something I've got to do. I can't imagine not diving; if only you would come with me once and look at my underwater world, see how magical it really is, you would understand.'

'But you promised,' he said. 'There might be another earthquake and you won't be as lucky as the last time.'

'Be patient, Darling, and stop worrying, I'm not taking any risks.' She kissed him lightly and picked up her silk drawstring pouch which matched her chungsam. The telephone rang just as they were about to leave the cottage and Mei-Ling turned back to answer it. She listened to the voice and a smile of delight lit up her face. She hung up and joined Young Chang in his car. They were going to pick up Old Crone Chang and take her to the masked ball given by the Capponcinis at the Club Royale. Young Chang had left his mother's house and moved in with Mei-Ling after their wedding. He had bought a piece of land abutting a small cove and they were building their new home there. It was close to the Capponcinis house but not as isolated, as it was accessible by a winding narrow road. They had engaged an architect who had taken all their wishes into account and the project was nearing completion. The small pier was already completed and both Young Chang's sailing yacht and Mei-Ling's dive boat were anchored there. The architect had given a moving-in date just after Semana Santa, and the Changs had already started packing.

'Who was on the phone? I suppose some more diving appointments.'

'No, no diving, my love. It was Dr. Johnson.'

Young Chang braked abruptly and stopped the car. 'Has it happened?'

'It's happened,' she laughed. 'it's truly happened.' He caught her to him and they held each other close.

'I think I'm going to cry with happiness,' he murmured.

They drove up to Old Crone Chang's house, and Mei-Ling got out of the car. 'I'll go, Darling. Your mother wants me to help her do her buttons up.' Mei-Ling walked through the front garden, knocked at the door and went in. Old Crone greeted her daughter-in-law lovingly. She thanked her for the orchids and Mei-Ling deftly did up the buttons on the old lady's black tunic.

'I'm glad to see you alone for a few minutes,' Old Crone said. 'I've heard of a new arrival on the island in whom you might be interested. He is staying at the Carlton and goes under the name of Muso Medusa.' Mei-Ling raised her eyebrows.

'Sounds like another Claudio Curria or Limbo Limone,' she said. 'hope that the Mafia isn't having another go at Tequila, Mother.'

'I thought you ought to know, Daughter.' Old Crone was aware that her daughter-in-law was working for the CIA, a fact that they had been able to keep from Young Chang, and her diving school provided a good cover for her work. 'It appears that this Muso Medusa has some connection with the young torero Caramba de la Cruz, who is due to participate in the bullfight this week.'

'I hope that we're wrong, Mother; I really don't want to be involved any longer. I thought we had dealt a death blow to them; besides, I'm going to hand in my resignation tomorrow. Do you want to guess why?'

Old Crone put her wizened old hand on Mei-Ling's smooth cheek. 'Dare I hope that my dearest wish is coming true?' Mei-Ling nodded and put her arms around the old lady's frail shoulders.

'God willing, Mother. In seven months time.'

In the presidential villa, the President Maria Beatriz and her husband Sam, alias Sorciero, were also getting ready for the masked ball. They had got married shortly after the earthquake which had taken the life of her brothers and changed the landscape of Tequila for ever. As the sole direct survivor of Big Daddy Del Rey, who had governed the Island since its independence, she had inevitably been elected President and acting Prime Minister of the small island republic. Her marriage to the local witch-doctor cum medicine man, had strengthened her position, and together they had begun a rule of true democracy. The revenue of the miraculously discovered oil field on Petrolia, the islet which had emerged from out of the sea in the wake of the earthquake, was distributed in disciplined and just fashion amongst the citizens of Tequila. Crime became a thing of the past, as no one felt the need to steal in order to feed hungry little mouths. The Mafia, which had established itself on the island with drug manufacturing and gambling, was ousted, and the only strictly enforced law was immigration. Applications for residence permits and acquisition of property were carefully scrutinized by the authorities, with the help of Interpol. Prosperity, law and order were the order of the day, and a sense of security prevailed.

Maria Beatriz and Sam said goodnight to the child in the nursery. The five year-old was a precocious, robust boy, full of spirit and laughter. He held out his arms to be picked up and hugged by his parents. They kissed the little boy and put him back into his bed. They left the nursery on tip-toes, and quietly closed the door. It was time for them to leave for the evening at the Club Royale. Maria Beatriz was resplendent in a new bullfighter's outfit which she had ordered in Spain. She had only missed one year in the arena after the birth of Young Charlie. She would be entering the ring as usual this year and fight, to the delight of the cheering public. Sorciero was wearing his feathered headdress and loincloth. He had thrown a mantle of leopard skin over his shoulders; the head of the animal was positioned on his chest. The glass eyes of the leopard looked ferociously at the world. The leopard skin had been handed down from his great-grandfather, the first official medicine man on the island. Sorciero looked impressive, the very epitome of a witchdoctor.

Usually Sorciero took the wheel of the Mercedes limo himself, but tonight he sat in the back with Maria Beatriz and let the driver do his job. Sorciero pushed a button, and the separating pane descended, giving them some privacy. Maria Beatriz looked surprised.

'What did you do that for, Sam?' she asked.

'I wanna talk about some things, before we get to the party, Sweetness.'

'Oh, oh, what's up? Whenever you start with that Sweetness stuff I get worried, Sam.'

'You're unfair, Sweetness, you've got nothin' to worry about. We got some information at the consulting sessions today, and Old Crone thought that you should know. A suspicious character's turned up in Tequila, according to one of our clients who works at the Carlton Hotel. He's called Muso Medusa and seems to have a connection to Caramba de la Cruz.'

'The bullfighter?'

'The same. Apparently they had some words at the Hotel, and then Caramba drove off with the Cap poncini girl, Miranda.'

'And there's another bit of business I want to discuss,' Sorciero went on.

'I knew that this was just the thin end of the wedge, Sam. What have you been up to, Lover?'

'Nothing at all. It's your advice I need about some business.' Sorciero sounded offended.

'O.K. Lover, you know that we can't do business in our position, not on the island anyway.

'Don't you even wanna know what it's about, Sweetness?'

'You know that the only thing you can charge for, is your herbal business and the occult seances.' She took his hand and squeezed it. 'If you need some cash just tell me, I have more than enough from my paychecks.'

'I wanna earn the money myself for your birthday present. I don't wanna give you a rooster and a basket of pineapples. You should listen to me, please. It is a kind of divining job and it pays very well. I've been offered thousands of dollars for a few days work and I said I would give my answer tonight at the ball. It's quite legal. Ain't you even a little curious, Sweetness?' He put his arm around her and nibbled her ear. She giggled like a young girl and snuggled up to him.

'O.K. Lover, shoot,' she cooed. Sorciero told her about the visit he had had that afternoon from Etienne Bonheur, the President of Mangorenia. She listened to the proposition and weighed it for a few minutes. They were approaching the turn off to the Club, and she knocked on the glass and motioned the young driver to stop.

'So Etienne is going to be at the party with Marthe?'

'Yea, that's when he wants the answer; tonight.'

'I don't know, Sam. Suppose you do find oil, won't it be direct competition to our oil in Petrolia?'

'He'll get some other diviner if I turn him down; my guess would be Haiti, Sweetness. If there's oil in Mangorenia, they'll find it, with or without me.'

'I think we should consult Old Crone; she'll be at the ball and we can talk to her before you give your answer. Agreed?'

Sorciero nodded and Maria Beatriz knocked on the glass and the young driver started up again. They drew up outside the entrance of the Club, and a bell boy opened the car door and helped the President out. Two uniformed policemen standing on the side of the steps saluted smartly. Maurizio and Domitilla Capponcini, resplendent in their French 18th century costumes, came down the steps and greeted the President and her husband, Sorciero. They embraced them and then escorted them into the club. The ball was in full swing. The band which had been flown in from Brazil was playing the gentle music of Carlos Jobim, Joao Gilberto and Stan Getz. A soft trade wind made the evening pleasant and guests were dancing on the polished dance floor. Waiters were making the round with drinks trays and waitresses in kimonos were passing around platters with Japanese delicacies. The costumes ranged from the opulent to the bizarre, and all the crème de la crème had gathered for this magical evening at the Club Royale.

The guests parted, leaving an avenue free for the President and her husband. The presidential couple stopped here and there to greet friends and dignitaries. The President was complimented on her exquisite outfit, and Sorciero nodded his head from left to right, shaking the feathers of the headdress.

'Did you spot the Bonheurs, Sam?'

'They are standing by the bar, Sweetness. They're about to start towards us. There are all the Changs, I've got to see them first,' he murmured to his wife. 'Let's head towards them so we can have a chat.' He picked up two glasses of champagne from a passing waitress and handed one to his wife. They approached the corner where Old Crone had settled herself on the high-backed chair. They all bowed to each other.

'We have to talk Old Crone,' Sorciero said. 'It's urgent. Where can we be undisturbed for a few minutes?'

'We can use the meditation room; the door is right behind us.' Old Crone said. 'Would you go and ask for the key at the desk and bring it here,' she asked her daughter-in-law.

The room was large, dark and cool. There were no windows, but a dim, soothing light was cast by special, partly concealed light fittings. There were a row of deck-beds set up and a white rug covered most of the floor space. It was unnaturally quiet.

'Can't we have some more light?' the President said.

'No, Ma'am that's all the light there is. You will notice that the room is completely soundproofed, so that no noise from the outside can disturb my meditation seances.'

'I didn't know that you had taken up seances here, Old Crone?' Sorciero sounded slightly aggrieved.

'They are not really seances, but classes like yoga, you know? It's part of the wellness package offered by the club.' Suddenly a strange haunting sound started, like the soft wail of an animal.

'Whatever's that?' the President asked. She shivered and took Sorciero's arm.

'Don't let that worry you, Ma'am, its just mood music, to put the clients into the right mood for meditation. It comes on automatically when someone enters the room. Well Sorciero, what do you want to talk about? Is it something to do with your visitor this afternoon?'

'You're right as usual, Old Crone. Etienne Bonheur wants to hire me to find oil on Mangorenia. They can't survive on Mangoreens alone. It seems that they've had a population explosion after the earthquake and they need another cash crop or something. They would pay me well. What do you say?' Old Crone padded up and down the silent room for an instant, mulling over Sorciero's words.

'The money is not important, Sorciero, although I can see that it would come in handy. Important is that you keep your finger on the pulse. They might get in some hoodlums who could somehow tap into 'Petrolia' if you don't go. I have a feeling that there might be another attempt by the Mafia to take over one of the islands, just like they tried to take over 'Tequila' some years ago.'

'So you think I should go.'

Old Crone nodded. 'Tell no one if you do find something. Come to me and we'll consult again and I will go with you to Mangorenia to take a look.' The pale lights suddenly went out, the music stopped and the room was steeped in complete darkness.

'Oh shit,' the President moaned, and clung to her husband. 'What's happened?'

'It's part of the meditation process,' Old Crone explained patiently.

'To hell with the meditation process; this place gives me the creeps. Where's the door, Old Crone?'

'We will have to wait a few minutes until the electricity comes on again. The door is automatically sealed.

'In other words, we're trapped here. No one can hear us and we can't get out'. Again Maria Beatriz shivered. 'You can count me out of this meditation lark, Old Crone.'

'The lights will be on again very soon, Ma'am. Please don't worry,' Old Crone tried to calm the agitated President.

'Do something, Sam; you're supposed to be the great witchdoctor. Make those lights go on again.'

'Yes, Sweetness, I'll try,' her husband murmured, and prayed for the electric current to return. He intoned an old incantation which had been taught him by his grandmother. To his amazement the lights went on, and the strange music started up again. Maria Beatriz gazed at him adoringly.

'You did it, Sammie, my love,' she crooned. Sorciero nodded silently, dumbfounded by the success of the incantation.

'Shall I accept?'

'Yes, Sorciero, accept,' Old Crone said softly.

'Now that's settled, can we leave this torture chamber and go back to the party?' Maria Beatriz said petulantly. 'I need another drink, and I want to dance to some good music.'

James de la Cruz stood at the edge of the large terrace and tried to locate the whereabouts of Miranda Capponcini. He had politely greeted his hosts on arrival from his cabana, accepted the champagne offered by the waiter, nibbled on a delicious little sushi, and nodded to people he didn't know, but who obviously knew who he was. He was undeniably handsome and elegant in his white Tuxedo, his silver blonde hair tied back neatly with a black ribbon. He was one of the very few who didn't wear fancy dress, having had no time to procure one. He breathed the balmy evening air perfumed by masses of gardenias, which had been placed all around the Club. Suddenly his nostrils twitched as he caught a different whiff of scent behind him, and he smiled. The hand which was placed on his shoulder was light and slender, and he turned his head and kissed it.

'Where have you been? You've left me alone for far too long, Randy.'

'I've been arranging our intimate dinner, as you commanded. Have you missed me?'

'That's the understatement of the year,' he said and turned to look at her. She was wearing a diaphanous cream coloured Sari, richly embroidered with silver and gold thread, which left her midriff bare. Her hair was piled high on her head, and her hazel eyes were glowing. Long, precious gold earrings were dangling from her small ears and a gold snake necklace was coiled around her throat.

'Who are you?' he whispered.

'I'm the Maharani of Slirinapoor, and I'm seeking my husband, the Maharaja. Have you seen him by any chance?'

'I wouldn't tell you if I had, Milady… tell me there is no such husband and that you belong only to me.'

'I belong to no one, Sir. Ah, he is over there… isn't he a fine figure of a man?' She gave a strangled little chuckle. 'By the entrance doors, trying to fight his way past the doormen.' James turned to see who it was, that had so amused Miranda.

'God, it's that Muso Medusa,' James growled. 'What's he doing here in this ridiculous get up?' Muso Medusa had spotted James, and was trying to catch his eye. Medusa had donned a multicoloured turban adorned by a feather held in place by huge a fake jewel. It sat askew on his head and was threatening to fall off his bald pate, as he struggled with the doormen.

'Looking for his maharani,' Miranda cooed. 'I think he's waving to you, James; are you going to rescue him?'

James de la Cruz turned back to her. 'Let's get out

of here, Randy. I don't know this man and he's embarrassing me,' he pleaded.

She smiled at him and took his arm. She led him off the terrace and they walked down the path leading to the waters edge.

'I'll save you this time, but you had better get rid of that ugly skeleton in your cupboard. It has a way of turning up too often.'

'I don't have a cupboard. Even if I had a skeleton, it wouldn't be in my cupboard.'

'Where would you be keeping it then? Under your bed?'

He laughed at the thought of Muso Medusa wedged under his bed. 'It would be a tight squeeze for him. Liable to be squashed for good. Let's forget this clown and go and have dinner. Where are you taking me?'

They had reached the jetty, and she got into the speed boat. 'Get in, and let yourself be surprised.'

'You're not intending to leave the bay, are you?'

'Afraid of the dark?' she teased.

'I don't know whether I'm a good sailor. It could be rough out there and I don't want to throw up.'

'Don't worry, we're on the leeside, and there's an almost full moon. Cast off and let's go.' She turned on the ignition and he did as he was bid. Miranda cut through the bay and slowed down to get through the passe. They were on the open sea now, but it was calm and the moon cast its silvery glow soothingly over the water. He looked for the stars and wished that he knew more about the constellations. The breeze brushed against his face and he felt a great feeling of well-being in his body. It started up in his toes and went right to the crown of his head.

'Miranda, this is so cool... I only once felt like that in all my life.' She smiled in the dark, but he could see her clearly in the moonlight.

'When was that, James?'

'The first time I saw you, Randy. I swear that's the truth.' He put his arm around her shoulders and held her close.

'You're an incurable romantic. Who would have guessed that the great Caramba de la Cruz, fighter of mad bulls, could have such sentimental feelings.'

'You can laugh at me, Randy, but I'm quite serious.' James gazed into the distance towards the horizon. 'What are those lights, over there?'

'That's Petrolia, but we won't go near there. There are coast-guards always controlling the perimeter, and we don't want to mess with them. We've almost arrived, and our dinner is waiting. Prepared especially by the best cook on the island, just for the two of us.' She drove the speed boat carefully into a small bay and stopped at a jetty which was illuminated by discreet spotlights. The stone steps up to the house on the cliff were also illuminated. James jumped out onto the jetty and helped Miranda up.

'Where are we, Randy? It looks marvelous.'

'This is a place which is reachable only by boat,' she said.

'They can't be doing much business, there's only one other boat here.'

'Did you really want company?'

'Only yours, my beautiful Maharani. Shall we go up?' He took her arm and they climbed the steps together, but not before she had discretely kicked the burglar alarm concealed in the second step. More lights went on. Whole garlands of fairy lights lit up the deck of the house. A table for two was laid in the middle of the deck, complete with lit candles, and frangipani blossoms scattered over the deep blue damask table cloth. The silver and crystal glittered in the candlelight, and on the side board a platter of small lobsters nestling on a bed of Russian salad, garnished with cherry tomatoes and arugola leaves, took pride of place. A silver sauce tureen filled with mayonnaise and lemon and lime wedges arranged around it, completed the display. A bottle of champagne was in the Old Sheffield wine cooler. Soft music came from the loudspeakers in the background. James de la Cruz took it all in and looked stunned.

'Welcome to "Tradewinds", our home in Tequila,' she said, and propelled him forward. 'Don't be shy, just make yourself at home, and open the bottle. I'm thirsty.'

Miranda Capponcini didn't lose her virginity after dinner that night, although she had fully expected to, indeed had waited for the moment with febrile anticipation. She had known from the very first instant that it was going to be James who would finally deflower her. She had resisted all attempts by a number of men of all ages who had tried to seduce her during the course of her adolescence. Her parents had allowed her considerable freedom, knowing that she was relatively pragmatic for her age, with her head well connected to her shoulders. They had warned her of the pitfalls of promiscuity, and the family doctor had instructed her on contraception. She had begun to take the pill, wondering when her prince charming would come along. Now he was here and her lips throbbed in unison with the rest of her body, as he kissed her gently at first, and held her to him.

'Oh, Baby,' he breathed, and stroked her breasts, feeling her nipples harden beneath the silk of her bodice.

'Oh, Baby,' she crooned and felt for the bulge in his trousers.

'You'll be sorry Randy, if you go on doing that. I'll tear that Sari to shreds.'

'I love you, James,' she whispered. 'It's ridiculous, I

know. We've only known each other for a day, but it's been such a wonderful day.'

'It has been for me too, Sweetheart. I don't know what I've done to deserve such luck. I'm overwhelmed by you.'

'Then overwhelm me and make love to me, Baby.'

He shook his head and released her. 'No, I have too much respect for you. My mother always told me that I must respect the girl I want to marry.'

'Oh, Baby, how sweet and old-fashioned, but can I really believe this? Perhaps you just don't find me attractive.' She pouted prettily.

'You're the most gorgeous girl I've ever seen and I'm crazy about you. You know that, Baby,'

'You'll have to prove that soon, Baby, but for now I'll let myself be respected. Let's get back to the party; as the hosts' daughter, I ought to be there to see the guests off.'

The last guests were sitting in the comfortable chairs in the bar area. They were deep in conversation with their hosts, and it was obviously a meeting of some importance. The American Consul, Jason Reed had taken off his Mickie Mouse mask, Sorciero had divested himself of his precious feathered headdress, and the Capponcinis had discarded their powdered wigs. The Young Changs had left early, but Old Crone was there, drinking a cup of steaming tea.

'There have been rumours that Curria's old crowd is preparing some kind of reprisals against 'Tequila' and the people they believe were responsible for his death and the failure of their business here on this island,' Jason Reed said. 'We've had a memo from the State Department.'

'So these crooks never give up,' Maurizio Capponcini cried.

'It may just be a rumour,' Maria Beatriz said hopefully. 'We check on all new arrivals.'

'Sometimes one rogue shark gets through the net. No security is ever fool proof. We have to be vigilant. The Semana Santa attracts so many tourists that it is difficult to screen everyone. It is the most dangerous time of year for 'Tequila'. I recommend you strengthen the coast guard around 'Petrolia', and double the scrutiny at the airport and the harbour,' Jason Reed said sombrely.

'There is a man who arrived on the island, an American with an Italian name,' Old Crone whispered. 'The bellboy Rodrigo, who came for a consultation this afternoon, told us the man is called Muso Medusa, and that he booked into the Carlton. He asked after Caramba de la Cruz. He said that he was a friend of the young bullfighter.'

'That nice young man? He couldn't possibly be one of them,' Domitilla said quietly. 'Randy's gone to have dinner with him at "Tradewinds" but she should be back any minute now. You don't think that she could be in any danger, Jason?' Domitilla jumped up from her seat and looked out to the water.

'There's no need to get so alarmed; it is after all, just conjecture.'

'I don't know what that word means,' Sorciero muttered, 'but I shall try and get an answer when I go into my trance tomorrow.'

'That's a great idea, Sorciero. I would also advise you, Mrs.President, to have one of your men to look through this Medusa's room at the Carlton and the young man's too.'

'The young man happens to have taken a cottage here. He didn't like the Carlton,' Maurizio said. 'I shall see to it myself.' He sighed and thought that Miranda would be distraught if it turned out that Caramba de la Cruz was involved in some nefarious plot against 'Tequila'. Maurizio decided to go through the cottage immediately, before the young people returned from 'Tradewinds.'

'I think we've discussed all the problems, and if you'll excuse me, Madame President, I'll go and give de la Cruz's cottage the once over immediately.' He bowed to Maria-Beatriz and Old Crone. 'You had better come with me, Jason; after all, you're the expert.'

Maurizio picked up a spare key to James's cottage and then he and Jason walked down the path which led to it. It was one o'clock in the morning and a light breeze whispered through the Casuarinas trees. It was a picture book view, with the moon reflected on the calm water of the bay.

'This place is so perfect, it's hard to imagine that anyone would wish to destroy it,' Jason said.

'Vendetta is instilled in Sicilians with their mother's milk. There's no escape,' Maurizio answered regretfully. 'I had hoped that we would be safe here, now.'

'You may still be, Maurizio. We have no definite proof.'

'Here's the cottage,' Maurizio fitted in the key and they went in. 'Let's leave the door open so that we can hear the boat.' He turned the light on. 'You look through the wardrobes, and I'll search the bedside drawers and the bathroom.' Jason nodded and opened the doors to the wardrobe.

'I don't even know what we are looking for, Jason.'

'Anything suspicious; documents, maps, letters, photographs, or something like this, for example.' Jason held out the revolver. He broke it open and extracted six bullets. 'Fully loaded and ready to go, Maurizio.'

'Is it illegal to have a weapon? I have a gun on board the cat.'

'I assume that you have a permit for it, and that you don't carry it with you when you take an aeroplane to

New York or Milano.' Maurizio shook his head. He was at a loss. Young Caramba looked such a nice young man; why would he smuggle a revolver into 'Tequila'.

'He may have a permit for it, Jason, and carry it legally from one country to another.'

'Only an Interpol man would have that privilege I think, and would have to contact the local authorities on arrival, etc. etc.'

'Maybe he's done that,' Maurizio said.

'That is a bit of wishful thinking, Maurizio. I would have heard of it from Chief Pereira.'

'He might have been too busy, what with Etienne Bonheur visiting and the security for the coming of the Semana Santa.' Maurizio seemed preoccupied and not keen on condemning the young bullfighter out of hand. Jason looked at his friend and decided to speak out.

'Maurizio, I can understand your reluctance to look at this revolver and to find excuses for its existence. I know that Miranda has taken the young man out to dinner. I also hear that they had lunch together at the Yacht club.'

'Are you tailing my daughter, by any chance?' Maurizio sounded aggrieved. Jason sighed and tried to pacify him.

'Mei-Ling told me that Miranda brought him to the club for lunch, and that they sat at the same table. The Changs thought he seemed a very nice young man, and that he was obviously very taken by Miranda.'

'Unfortunately, Miranda is very taken with him too. I'll have to tell her to keep her distance until we find out more about him.'

Jason looked thoughtfully at Maurizio. 'She's quite a headstrong girl, your Miranda. It might push her into exactly the opposite direction.'

Maurizio cocked his head and held his hand to his lips. They heard the swish of the powerboat in the water and the hum of its engine.

'Let's put everything back and get out of here. I think they've returned.'

James wandered down the path whistling under his breath. It was not a loud triumphant tune, like the victors march from 'Aida'. Rather it was a quiet sentimental 'It had to be you'. He felt strangely exhilarated, proud of having held back and preserved the virtue of his future wife. He felt so full of love for Miranda, that it hurt somewhere in the region of his chest. He unlocked the door to the cottage and looked around him with satisfaction. The room was pleasantly cool. The king size bed was neatly turned down, and his pyjamas laid out on it. His slippers were placed side by side by the foot of the bed, and fresh hibiscus blooms were strewn over the coverlet. A miniature bottle of brandy and some chocolates lay on the pillow. The ice bucket was full, and fresh mineral water and glasses were on the tray. James poured out some sparkling water, and got out of his clothes. He would take a good shower before slipping between the fragrant linen sheets. He turned on the tap and waited until the shower spurted ice cold water. He gasped as he stood under it and then hurriedly put on the fluffy bathrobe. James studied his face in the mirror above the basin and grinned at his reflection. He looked absurdly pleased with himself. He finished brushing his teeth and dried his hair with the dryer which hung on the wall. Now he was ready for bed. It had been a long day and there was not much of the night left. He pushed open the door and stared at the unexpected visitor sitting in the armchair.

'What the hell are you doing here? I can't remember having invited you to visit,' James shouted.

'Keep your voice down, boy. You don't want the security guys to come in, do you?'

'You're damned right I do. I'll call them now,' James said and went for the phone. Muso Medusa was a fast mover in spite of the overweight he was carrying, and reached the instrument before James could get hold of it.

'You're beginning to piss me off, boy. Suppose you calm down and listen to me.' Muso Medusa's tone had changed considerably. It was no longer friendly and jovial. There was a definitely steely note in his voice now. 'Sit down and shut up.' Medusa unceremoniously pushed the protesting James into the armchair.

'This is the last warning, you little punk. You don't really think the family has wasted all this money on you and your mother, for you to turn your back on it now. You have a duty to fulfil the vendetta. You have been given a heaven-sent opportunity to do it. So you don't want to know me, that's OK, but remember that you have a week to deliver.'

'And what if I don't?'

Medusa smiled thinly. 'You will be exposed for who you really are; a member of the Curria clan. You will most probably be thrown into jail on this godforsaken island and left to rot here for years. Furthermore, your pathetic idiot of a mother will suffer dire consequences. She will probably disappear without a trace. Not that anyone has ever tried to trace her, as far as we know. She's just another dumb American girl who trapped your father into a marriage because of you.'

James sprang up and pummelled Medusa on the chest. 'Don't you dare call my mother names. Leave her out of this.'

'We can't leave her out of it, sonny,' Medusa smirked, and retreated towards the door. 'You don't deliver, and you'll never see her again.'

'You've had your say, so get out of my room,' James growled and clenched his fists. The door closed behind Muso Medusa and James threw himself onto the bed. He had faced a number of irate bulls in the arena and a number of irate fathers whose daughters he had led astray, with equanimity and courage. He had turned countless veronicas and had met the bull's fiery eyes bravely, without flinching, as it stampeded past him. He had pierced the bulls necks with the blade hidden underneath the mulleta. He had even been challenged to a duel, and had acquitted himself honourably. Now, for the first time in his young life, he felt totally impotent. It was a feeling he had never experienced before and he didn't like it, didn't know how to cope with it. Tears of frustration started to form in his eyes, and that was the ultimate humiliation. He vowed that he would find a way out of this impasse, and turned onto his side, hugging the extra pillow. He fell asleep instantly, like very young children do, breathing evenly and quietly, until the first rays of sunshine woke him in the morning.

The same rays wandered over Maurizio and Domitilla's double berth on the cat. She stretched her long limbs and he let his hand linger on her hip. The cat swung gently around its anchor, and the tackle whirred in the light breeze. She murmured something unintelligible and put her hand over his.

'I love you too,' he said.
'Can I smell coffee?'
'No, that's just wishful thinking, my love.'
'Can you make my wish come true?' She wrinkled her nose. 'I'm sure I can smell coffee. Miranda must have put it on.

He swung his legs over the edge of the bed and sighed. Domitilla opened her eyes and looked at his hunched back. 'You must be tired,' she said. 'You tossed and turned most of the night.'

'I'm sorry, my love; I tried not to disturb you.'

'You're worrying about Randy and the young man, I know. I'm worried too. I think we must tell her to distance herself a little.'

'That might just push her in the wrong direction. You know how headstrong she is,' Maurizio objected.

'Nevertheless, she is a sensible girl, and we've always discussed things with her. We must tell her of our concern, the sooner the better.' He nodded and got up.

'I think I'll go for the coffee now. I can smell it too.'

'Don't bring it down; it looks like such a beautiful morning. Let's have it on deck with Miranda.'

Maurizio got up and went to the head. Domitilla followed him and they stood brushing their teeth at the basin. She brushed her recalcitrant long hair, trying to untangle the strands. She splashed some cold water over her face and neck and briskly dried herself.

'I suppose I had better put on my T-shirt. You never know who's going to pop up from the sea.' She slipped it on and Maurizio looked at her and grinned.

'Don't get me wrong, but you look better without.'

'Flatterer,' she pouted.

'I think it's time we spent a night at "Tradewinds". I'm feeling deprived.' He reached for her and brushed her breasts lightly until her nipples raised the shirt. He lifted it and cupped her generous breasts in his hands. He bent his head and took first one, and then the other dark pink nipple in his mouth, caressing them with his tongue. She leant against him, weak-kneed.

'God, even after all these years...' he murmured, 'you're still the most exciting, sexy woman. I want to fuck you now.'

'Stop, Maurizio, pas devant les enfants.'

'Coffee's ready, and I'm going for a swim,' Miranda

called from deck. 'I won't be long.' They heard the splash as their daughter dived into the water. Maurizio slowly peeled off Domitilla's shirt, and lifted her onto the bed. He took off his shorts and she reached for his erection.

'Not a bad way to start the day,' he said and let his mouth roam over her body.

The sun beat strongly against the boat now, and Domitilla stirred in Maurizio's arms. She looked at her wrist watch and sat up with a jolt.

'Wake up, Darling. It's gone nine o'clock. The coffee must be cold by now.'

'Not if it's sitting in the sun. Besides, it's a small price to pay for...' She put her hand over his mouth.

'Yes, isn't it?' She got up and smiled. 'Let's have a shower and get ready.'

Fifteen minutes later they came up top and found the breakfast that Miranda had prepared. She had poured the coffee into a flask to keep hot, and had put out croissants and tiny brioches. Brightly coloured mugs and paper napkins were placed on the teak table. Of Miranda, there was no sign.

'Where is the child? The dinghy's here, so she didn't go ashore,' Domitilla said anxiously. Maurizio picked up the binoculars and scanned the coastline.

'Just as I thought; she's sitting on the rock with young de la Cruz.' He started to wave, but couldn't attract their attention. They were entwined in each other's arms and had seemingly abandoned the outside world. Maurizio cursed under his breath. He went below and came back with the bullhorn.

'Miranda...' it boomed across the bay. Maurizio saw the young lovers start apart and he waved again and got an answering wave from Miranda. 'Come back, Miranda,' it boomed again, and he had the satisfaction of seeing Miranda dive neatly into the water and headed towards the catamaran. She reached the side and climbed the steps.

'Go and take these wet things off and come and have breakfast, Randy.'

'Yes, Papa,' she answered.

'We have to talk, Sweetheart.'

'Yes, Mama,' she beamed. 'I can't wait to tell you.

A few minutes later she was back, wrapped in a white bathrobe, and looking extremely pleased with herself. She took the coffee her mother offered her, and wolfed a small brioche.

'Lord, swimming does make one hungry,' she said and sat down between her parents.

They looked at her lovingly, and Maurizio cleared his throat.

'Sweetheart,' he said slowly, 'your mother and I are worried about you; more precisely about the way this new romance of yours has progressed so rapidly.'

'Oh, Papa, stop worrying; your little girl is still a virgin. James respects me too much and wants to wait until we are married.'

'Married?' Domitilla and Maurizio cried simultaneously.

'Yes, he asked me to marry him. Isn't he wonderful?' Miranda positively glowed with pleasure.

'That's totally out of the question, Sweetheart,' Maurizio said decisively. 'You're much too young to get married.'

'Mama wasn't too young at nineteen to marry you, Papa.'

'That was different; we knew each for some time and were engaged for six months before we got married. We knew each other's families, came from the same background.'

'Don't be so old-fashioned, Papa.'

Maurizio bridled. 'Old-fashioned? Me? Old-fashioned? Why, we have been the most liberal parents in the whole world,' he said.

'OK, I admit that you are both usually really cool,' Miranda beamed. 'He's coming over to "Tradewinds" for lunch, so that he can formally ask for my hand in marriage. Isn't that sweet and old-fashioned enough for you?' Maurizio and Domitilla looked at each other in despair and Domitilla stroked her daughter's cheek lovingly.

'Yes, very sweet, but what's the rush. You're in the middle of your studies, and we want you to get your degree, before you take such an important step. Besides, we hardly know anything about this young man, except that he is a fine bullfighter and is good-looking.'

'Oh, Mama, he's beautiful, and I love him.'

'That's all very well, but we must know more about him; who his parents are, what kind of education he's had, whether he can afford to keep you. His is a very dangerous profession. What if he gets disabled, or has to retire, and can't fight anymore? Will he have a pension? How will you pay for your children's education?'

'I've thought about that, and no sweat. You'll give him a job in your company. We'll be great at running one of the clubs. You're training me already for that, aren't you?'

'We hate to dampen your enthusiasm, Randy, but there have been rumours,' Domitilla said.

'Oh, I know; I've seen the "Hello" magazines. I know all about his peccadilloes.'

Again Maurizio and Domitilla glanced at each other.

'You tell her, amore,' Domitilla murmured.

'It's not the peccadilloes, Sweetheart. It's good for the man to have some experience. We have reason to believe that he's somehow tied up with the Mafia.'

Miranda looked wide-eyed with astonishment. 'You can't be serious! All of you on the island are quite paranoid about the Mafia plotting to return and wreak havoc on the island.'

'Don't pooh, pooh the possibility, Randy; the threat is very real. Your young man is carrying a gun.'

'Good Lord, is that all? He showed it to me. He was told that there was a lot of crime on the island and that he should carry it for his own safety and protection. I told him that it was absurd, that the island was safe as houses, and that he had been totally misinformed.'

'What about this strange fellow who is following him around everywhere?'

'James doesn't know the guy. He never saw him before in his life.'

'The guy certainly says he knows him.'

'I choose to believe James. Mother, we are in love,' Randy squinted at Domitilla. 'You understand that, don't you? You always told me that the same thing happened when you met Papa. Love at first sight.' Maurizio took his daughter's hand and looked into her bright eyes.

'Tesoro, don't be in such a hurry. What do we know of this young man? We would be irresponsible parents if we didn't look into the background of this boy before according him your hand in marriage, as he puts it. What's his real name anyway?'

'His name is James Harvey, and I'm sure his family is perfectly respectable,' Miranda said and tossed her mane rebelliously.

'We'll discuss it when we get to "Tradewinds". Haul in the anchor, Randy, and we'll be on our way,' Maurizio said calmly went to the helm of the cat and started the powerful engine.

Chapter Three

Sorciero was sitting in the saloon of Mangorenia's only warship, an antiquated Hovercraft which had been adapted to take the presidential limousine on board. It also doubled as a coast guard vessel, and sported a machine gun on a turret. The old lady's speed was steady and slow, but since she had never had to give chase, it was a matter of no importance. She hovered over the sea at a steady pace, rocking gently from side to side. Sorciero saw the outline of Mangorenia slowly take shape. The extinct Volcano rose against the sky, dark and somewhat threatening. It was barren of vegetation apart from the Mangoreen trees which neatly crisscrossed its flanks. At sea level there were patches of greenery, thanks to the desalination plant which gave the island enough water to grow its cash crop and the necessary vegetables and fruit. The harbour was too small to take any big ships, and the container vessel which stopped there once every four weeks, had to anchor off shore and unload its cargo onto a couple of tenders. The passe into the harbour was narrow and suitable only for pleasure craft, and the handful of fishing boats which provided the island with fresh fish. The presidential hovercraft skimmed over the reef and settled onto the concrete platform on shore. A uniformed sailor sprang to open the hatch and stood to attention.

'Well, Sam, our eagle has landed,' Etienne Bonheur

said jovially. 'Shall we disembark?' He turned to his wife. 'After you, my dear.'

Sorciero followed the presidential couple down the rickety steps, and into a small concrete structure which was the departure and arrival lounge of the miniscule airport. There were some chairs and tables along the walls, a noisy old air-conditioning unit, and large cooler with an inscription on its side. It read 'two litres of fresh mangorenia water a day, will keep the doctor away.' The customs and immigration officer sitting in his cubby-hole sprang to attention as the presidential party walked through. Then he raced to the cooler and took out some bottles of cold water which he offered to the party.

'Here Sam, you must taste our water, it's really good,' Etienne Bonheur said and handed Sorciero a bottle. Sorciero nodded his thanks and took a sip. He made a small grimace, then took another sip.

'Well, what do you think of it?'

'It's not as fizzy as Perrier, Etienne. It tastes strange at first, but then it grows on you.'

'That's an excellent definition of it. Our people love it more than coca cola, would you believe?' Etienne Bonheur smiled happily. 'Come along now, Sam. We'll drive home and you can unpack your bag, have a little breakfast, and start work.'

The white limo snaked along the winding road which led up to the foot of the Volcano. There, in the shade of a copse of old Mangoreen trees, was an oasis of flowering gardens and splashing fountains, which surrounded the presidential villa. It was of course considerably more modest than the pink marble-clad presidential villa in Tequila, but it had that certain charm of an old colonial Great House. It was comfortable and cool. Sorciero unpacked his small bag and tested the king-size bed. There were some chocolates on a tray. A bottle of water and a cut glass tumbler were also on the bedside table. Sorciero unscrewed the top and put the bottle to his mouth, ignoring the glass tumbler. Again he felt a strange tingling in his mouth as he swallowed. It had to be the desalination plant, he thought, which had gone on the blink. He prodded his belly, but felt nothing. Hopefully he wouldn't get an upset stomach. He decided to close his eyes for a moment, and drifted into sleep. The brisk knock on his door made him rear up, and stare around him in confusion. The door opened, and a pregnant young woman came into the room. Her bosom was bursting out of her blouse and Sorciero felt a sudden erection as he looked at her. A heavily pregnant woman had never aroused him so before, and he felt more confused than ever.

'Sorry to disturb you, Sorciero, but your jeep is ready and waiting, I'm to tell you,' the young woman said. 'Also I'm to ask you whether you need anything, like some food and drink to take along.'

'Yea great, that will be fine. Perhaps some local beer and a sandwich.' The girl nodded and left the room. Sorciero put his hand on his crotch and was sorely tempted to indulge in a bout of masturbation. He decided to take quick cold shower instead, and within 15 minutes was standing by the jeep, ready to go. He checked the equipment stashed in the back of the vehicle. He had asked for a bucket and spade, and had brought his divining rods with him. One was a primitive one, fashioned out of a forked branch. The other one, made out of light metal, had been a gift from Maria Beatriz. He had taken it along to please her, but preferred to use the forked branch, which he had whittled down to the right thickness. A young man was sitting in the driver's seat. He was the designated driver and guide and introduced himself as Mark.

They drove along the tarmac road which circled the island at the foot of the Volcano. Sorciero was studying the map on his lap and pursed his lips. He couldn't for the life of him imagine where to begin his search. He didn't have enough time to do a systematic search, so it had to be a random one. Common sense told him that oil wouldn't be in the volcano, or on the slopes. He sighed and studied the map again. He would start by the salt water lagoon which was close to the small harbour.

'I want to go to the lagoon, Mark.'

'OK Boss, but there ain't much there except thousands of birds. Some folks say they have seen alligators there, but they was very drunk at the time. The folks, not the alligators.'

'Alligators? That's strange, we've never had anything like that on Tequila.'

'Folks say that a fellow had a pair as pets on his farm, and that they escaped.'

'Never mind, Mun, let's go to the lagoon.'

They drove along the coastal road through the plantations of Mangoreens. Men and woman were working in the orchards, pruning, and hoeing around the slender trunks. The sun was hot and the air dry. Sorciero idly watched the landscape go by. There weren't many vehicles on the road; cars were expensive and bicycles were the preferred mode of transport. Now and then, he saw a donkey in the fields, carrying baskets laden with Mangoreens. Two women were ambling down the road, chewing on the fruit. They were both pregnant.

'Hey, Mark, you folks sure like to fuck; and you sure knock them fillies up.'

Mark flashed a happy smile at Sorciero.

'Yea, Mun, our birth rate is way up since the

earthquake. There ain't much else to do, since the earthquake knocked out the radio and TV station, see?'

'Why don't they fix them things?'

'Not enough cash around, the government says. It's ploughing it all back into the Mangoreens. Besides, we likes to fuck. You know, Mun, I can keep an erection going for a long time. Makes the fillies bawl.' Mark's smile got even wider, and he swerved around the bend with true panache. 'Here's the lagoon, Sorciero. Where do you want to start?'

'Stop over there, Mun. There's some shade by the fruit vendor.'

Mark rolled towards the fruit stand and stopped the jeep under a thatched parking bay. Sorciero got out and walked to the edge of the water. The lagoon lay still; a large stretch of calm golden water, which reflected the sunlight. Long-legged Heron were wading at the edge hunting for their food, arching their long necks, waiting to pounce on their prey. Little black, red-capped moorhens floated by with their progeny. High above him, frigate birds were cruising gracefully in the sky.

Sorciero walked back to the jeep and took out his divining rods. He stopped by the fruit stand and bought a couple of Mangoreens. The woman smiled at him happily as she pocketed the coins. She came out from behind the stand and Sorciero saw that she was pregnant. The bitch that waddled behind her was in the same condition.

'Nice lagoon we have here, Mun,' she said. 'You stranger to the island? Lots of tourists come to see the lagoon.'

'How come every one of you ladies I sees here in Mangorenia is in the family way?'

'I guess God has blessed us with more fruitfulness

than other folks, Mun.'

'Yea, your President said something like that.' Sorciero grinned at her and suddenly felt a stirring in his groin. He turned away and beckoned Mark to follow him with the spade and shovel. He put on his old straw hat and gave a small wave, to hide his embarrassment. He couldn't understand his reaction to the fruit-vendor. In fact, in the past, he had found women rather unattractive in the later stages of pregnancy. He walked off into the fields holding his rod steadily in front of him. There was no response as he walked up and down concentrating on the job on hand. Now and then he stopped and wiped his forehead, and took a drink out of the water bottle. He really would have to talk to Etienne about the quality of the water. Probably the maintenance of the desalination plant had been sadly neglected. Sorciero continued the pattern until he had reached the beach. Suddenly the rod responded violently, revolving in his strong hands. He dropped the rod, and said to Mark:

'Dig here, Mun and let's see what gives.'

Mark dug in with the spade and lifted the black sand. Sorciero sat back on his heels and sifted the sand earth through his fingers. He felt sure that the rod had reacted to metal, not to oil. He had experimented on several occasions with his divining rod and could tell when it was responding to metal, be it gold, silver or iron. The violence of the reaction pointed to gold.

'How far down should I dig, Boss,' the driver wanted to know.

'Until you find something,' Sorciero muttered.

'It must be at least three foot down already. What do you expect to find? A pot o'gold?'

'Maybe, maybe... Here, let me have a go.' Sorciero

took the spade and continued to deepen the hole.

'Now it's almost big enough to bury someone,' Mark said.

'You're right, Mun, something was surely buried right here.' The spade hit something solid, and both men stared into the cavity. 'Let's clear the earth away quickly, and see what it can be.'

Mark and Sorciero worked feverishly until they had laid bare a long wooden box.

'This looks just like a coffin, Boss,' Mark said softly. His voice trembled slightly and a shiver ran over him. 'It's 'mal chance' to interfere with dead; I'm out of here.'

'It's not a coffin; besides, even if it were, it can't harm us. I will sing an incantation against the mala suerte.' Sorciero stood up and lifted his arms towards heaven. He chanted the appropriate incantation, and hoped that Mark would be suitably impressed.

'Right, Mun, let's go for it. The thing has handles, so let's pull it out.' Mark reluctantly tugged at the casket, then let the handle drop.

'Why can't we just take the lid off without moving the box,' he whispered.

'OK, then; let's do that.'

Sorciero prised open the lid with difficulty, using the edge of the spade. They were greeted by a grinning skull and they both sprang back.

'I told you, a bloody coffin, Boss,' Mark whimpered. 'Not a pot of gold. Let's just cover all this up again;

Sorciero shook his head and peered over the edge again. The skull was still grinning, showing all its teeth. They were shiny and yellow; a perfectly executed denture made out of metal. There was no doubt in Sorciero's mind that they were cast in gold. A gold, stone encrusted earring lay on the side, and a leather pouch. There were remains of cloth on the bones, and a wooden stump proclaimed that the body only had had one leg at the time of burial. Sorciero straightened up and looked at Mark, who was still shaking from shock.

'I wasn't wrong, you see? There's gold down there, Mark.'

'Don't ask me to pull those teeth. I ain't no dentist. Besides, I've told you before, it's 'mal chance' to interfere with the dead. How long do you think the poor bastard's been down there?'

'Dunno, for sure. Perhaps a couple of hundred years. Must be some pirate captain. Do you know of any pirate ship sunk in this ocean?'

'They says that 'One-Stump-Matelot's' ship ran aground in these parts.'

'Who's they?'

'My granny used to tell us the story. He sure was one crazy fellow.'

'And how would your granny know about that?' Sorciero sounded doubtful.

'Because her granny told her. Her granny was connected to him.' Mark fished around under his T-shirt.

'Look,' he said and pulled out a gold chain on which hung a gold coin. 'Old 'One Stump Mat' gave this to her.'

Sorciero fingered the coin and looked at the markings. It certainly looked like gold and the coin seemed ancient enough. He picked up the pouch, and shook out the contents. A mass of gold coins fell out. They were identical to the one Mark had on the gold chain. Sorciero glanced at Mark and then at the skull.

'There's a distinct family resemblance,' Sorciero cackled. Mark smirked and took another look at the skeleton. 'Can I have the gold teach and all?'

'How about the 'mal chance' bit?'

'It don't matter if it's a relative,' Mark said piously.

'It's nothing to do with me, Mun. This here gold

won't make any difference to Mangorenia's economy. Let's cover it up, and you can mark the spot. What happens after that is none of my business.'

Sorciero continued his exploration for the next two hours, and turned up nothing. Not even a copper coin. They returned to the lagoon and Mark took out the cool box which contained their lunch. They sat under the thatch of the fruit stand, which had been abandoned by the seller. It was lunch and siesta time for Mangorenia, and nothing stirred in the heat of the noon sun. The birds had taken shelter in the mangroves and the lagoon lay still; not even a ripple disturbed the clear water. Sorciero took off his clothes and dipped into the water. It was just the right temperature, and felt like silk on his skin. He waded back to the shore and returned to the fruit stand. He pulled on his jeans and his shirt and smiled at Mark contentedly.

'Nice water you got here; I remember coming to this island with my old man when I was a child. Nothing has changed much since then.'

'I never been off island,' Mark said. 'I bet you've been many places, Sorciero.'

'I've been to New York. What a place, Mun.' Sorciero shook his head. 'It's scary; all these huge skyscrapers, thousands of cars...

'And all them gals; did you? you know?'.

Sorciero managed to look shocked. 'I'm a married man, Mark.'

'So am I,' Mark looked sheepish. 'But all them cunts...enough to make a man wild.'

'You people don't seem to be hard up for a fuck. I've never seen so many pregnant women in all my life.'

'We Mangorenians have pretty good sperms, Sorciero. Those little suckers sure know the way to a woman's heart.'

'So that's what you call a woman's heart in these parts?'

'When we'se being polite and gentlemanly like, yes, Mun.'

'Your woman pregnant as well?'

'She's the most pregnant of them all,' Mark said proudly.' Her belly so big, she can't get through the door almost.'

Sorciero and Mark drove further along the coast, stopping now and then for Sorciero to examine the sandy earth more closely. They had almost circumvented the island, when Sorciero asked Mark to stop again. The ground looked different. The black earth had become white, almost crystalline. A deep cut into the volcano's side revealed a white stone formation. Some men were carving the stone out of the cliff and loading it onto a van.

'What's this then?' Sorciero asked curiously.

'It's our stone quarry. We build our houses with this stone. Didn't you notice all the white houses? We ain't got enough wood to build houses like in other places.'

'That's cool, Mun. Stone is better than wood. Keeps you comfortable in the heat. OK. I'll take a couple of minutes around here.' Sorciero got out of the jeep and walked along the road into the quarry, holding his divining rod in front of him. The men looked at him strangely, then continued with their work. Sorciero bent down and scooped up some of the debris left by the stone cutters. It felt gritty between his fingers. The crystals shone in the fading sunshine. It was nothing like the lime stone quarried in 'Tequila'. Sorciero walked on, but his divining rod remained unresponsive. He returned to the jeep and jumped in.

'Home, Boss?' Mark inquired.

Sorciero nodded and looked at the last rays of the

setting sun over the ocean. It was a glorious horizon, with just a few tiny clouds framed in gold hovering around the bright orange orb. The jeep rumbled over the potholes and finally came to the cross roads which led to the Presidential Villa. A slight breeze started to rustle through the Mangoreen and the Casuarina trees that lined the road to the Great House. The heavy perfume of Oleanders hung in the air, and the soothing spray of the fountain greeted Sorciero as they drew up in front of the screened doors. He got out and took his divining rods under his arm. It was time for a shower and a drink and some real food. He nodded to Mark, and entered the Great House.

Dinner was served in the dining hall at 7.30 precisely. There were a handful of guests attending and they filed into the hall after the President and the first Lady. Sorciero, the guest of honour, was seated beside the first Lady, and looked crisp and cool in his Ralph Lauren linen suit. The first course was a Mangoranian cocktail, liberally laced with rum, lime and brown sugar. The main course was a huge dish of fresh caught Lobster bedded on fried seaweed, accompanied by a butter sauce. Crusty French bread was served with it. It was baked in the Great House kitchen daily. Sorciero remembered not to lick his fingers, but used the fingerbowl set before each guest. Maria Beatriz would have been justly proud of him! A truly French cheese board garnished with purple grapes rounded off the meal. Etienne Bonheur looked around his highly polished mahogany table and was pleased. Madame Bonheur rose and asked the ladies to accompany her into the drawing room.

Etienne Bonheur dismissed the butler, and the gentlemen were asked to pour their port and light their cigars. A sigh of contentment seemed to rise from them all as they puffed on the Cuban cigars, and sipped the wine. Etienne Bonheur looked expectantly at Sorciero, and asked.

'Well, my friend, what can you tell us about your day? We are all very curious to hear if you've made progress.'

Sorciero shook his head regretfully. 'No progress, gentlemen. I've covered some ground along the lagoon and along the coastal road, but nothing moved.'

'Not very encouraging,' the Minister of Public works murmured.

'I'll go out again tomorrow,' Sorciero said. 'You mustn't give up hope so quickly.' He looked around the table with a bright smile. He hoped that they wouldn't ask him to pack it in already. 'Besides, I see that you have a good quarry for stone. It looks very like marble. Why don't you try and export that?'

'It is only a small quarry, and we use all the stone for our housing,' the Minister of Public works said firmly. 'What we need is oil, just like you have, Sam. Find it for us.'

'I'll try, believe me, Mun,' Sorciero said earnestly. 'Can I say something more? Your water tastes strange. Is there something wrong with the desalination plant?'

'But not at all, mon brave. It's working perfectly fine. The water you are drinking is spring water.'

'Spring water?' Sorciero sounded amazed. 'Since when do you have spring water? I thought you got all your water from the desalination plant, and the odd drop of rain, if it ever falls in these parts.'

'We discovered a spring after the earthquake. It's lightly sparkling and sometimes a trifle salty, but we got used to it and like it. At least we don't have to import Perrier from our ex Mother country any longer.' Etienne Bonheur looked around the table and puffed on his cigar. 'We can give you a few more days,

Sam. We will all hope and pray that you'll come up with something. We have run out of ideas and only a miracle will save our island from total bankruptcy.'

'You folks seem to be fucking like rabbits. It seems to me that your next project should be a birth control clinic,' Sorciero said and drained his glass. Etienne Bonheur looked at his guest of honour reprovingly. He knocked off the ash of his cigar and rose. 'Gentlemen, shall we join the ladies in the drawing-room for coffee? And remember Sam, that there are ladies present there.'

Sorciero woke up next morning and stared into the figure of the maid who held his morning coffee on a small tray. The sight of her made his penis spring into action with such force that he gasped.

'Your coffee, Sorciero,' she smiled brightly at him.

'Thank you. Just put it down somewhere, please.'

She turned, and the sight of her ample buttocks moving under her lightweight uniform, pitched his arousal to almost breaking point. It took all his self-control not to grasp her, tray and all, and drag her into bed with him. She put the coffee down, and turned back to him.

'Can I ask you a favour, Sorciero? It is said that you have powerful magic.'

Sorciero cleared his throat and looked down his nose.

'Well what is it you want?' he muttered.

'It is said that you can tell whether it will be a boy or a girl, just by putting your hand on a woman's belly.'

'And that's what you want me to do? Put my hand on your belly?' The girl nodded enthusiastically and came as close as she could to the bed. She thrust out her swollen belly, and patted it. 'Touch me right here, Sorciero; that's where it kicks.'

'I can't do that, girl, you've been told wrong,' he said.

'I can pay you, Sorciero. I've got two Mangoreen francs. Please do it, so I can tell my fiance'. He says we get married if it's a boy.'

Sorciero groaned and covered his face with one hand. With the other he stroked the proffered body. He felt an electric shock flash through his arm, and he winced in discomfort.

'What was that?' the girl whispered and stepped back.

'That, girl, is magic,' Sorciero said proudly and lay back on the pillows. 'There are twins in this big belly of yours, girl. A boy child and a girl child.'

'Oh, Mun, that's great! Wait till I tell my fiance'. Thanks, Sorciero.' The maid fairly danced out of the room, leaving a panting Sorciero on the bed. His erection subsided slowly, and he picked up the telephone and asked the operator to put him through to the Presidential Villa in Tequila.

'Hallo Sweetness,' he bellowed down the phone. 'Can you hear me? Good, good. Nothing to report on that front, but listen to this, Darlin; I can tell whether it's a boy or a girl; just by putting my hand on a belly.' He held the receiver away from his ear. 'Stop yelling, Darlin. The belly was fully clothed. I've never done this before, except with you. It works on other folks as well. I'll be able to charge at least three dollars for that at home. How's the boy? Yea, Yea, I miss you both too. A lot. You don't now how much, Sweetness.' Sorciero planted several kisses onto the receiver and hung up.

The jeep rode past the lagoon, and Sorciero looked at the water with longing. It was hot and dry, and he was thirsty. Mark was driving him to the opposite side of the island where the Atlantic beat against the rocks, throwing up giant sprays of water. They stopped at sign which read 'Dear Tourist, stop here and admire the devil's maw.' Mark walked along the slippery rocks, Sorciero in his wake.

'There it is, Mun. The maw.' They both gazed into the deep chasm where the water seemed to boil. 'Every seven minutes, the maw spits out the water. Them stupid tourists don't know that and get sopping wet,' Mark said and looked at his watch. 'It is said that if the maw spits out a fish it is 'bonne chance'. Stand back now, Mun.' He pulled Sorciero back from the edge of the maw, and they heard the increased noise of rushing water. It reared up, exploded high above their heads, and showered down on to the rocks. A silvery brown grouper landed at their feet, flopping about helplessly.

'I knew it, Mun; I knew it would happen if I brought you here.' He held up the wriggling fish up by the tail. 'You're going to bring us 'bonne chance' Sorciero.'

'Not much of a catch, Mark. How can a miserable fish help Mangorenia?'

'It's not the fish, Mun, although we could put it on the barbecue for our lunch. It's the omen that counts.'

Sorciero nodded and turned to go back to the jeep. 'I wonder whether the fish is pregnant as well,' he muttered as he got into the jeep. Mark put the grouper into the cooler and got out a couple of bottles of Mangoreen juice. He offered one to Sorciero and they drank in silence, enjoying the full blast of the air-conditioning.

They drove for another 15 kms then Mark pulled off the road, and parked the jeep under a Casuarina tree. They were quite close to the sheer cliffs that rose on that side of the Volcano. Mark got out and took the cooler to the entrance of a cave in the rock. It was shady and cool at the entrance, and Mark brought out a couple of folding chairs and some kindling. He built a fire and put a rusty old wire grill on the flames. He unpacked the sandwiches and the thermos of iced tea. Then he started scaling and cleaning the grouper. Sorciero took up his divining rod and entered the cave. Not that he expected to find anything in there, but after all, he was being paid for doing a job and he was conscientious about his work. The entrance was wide and shed enough light for Sorciero to walk a little way in. His rod gave no sign of life as he crisscrossed further into the cave. A shout from Mark got him running back to the entrance.

'What's up Mun? A rattler got ya?'

'Ain't no snakes on this island, Sorciero,' Mark laughed. 'But lookie here what I found in that ol Grouper. It was pregnant alright. Right in its belly it was.' He held up a ring set with a blue and a white stone. The white stone sparkled in the sunshine; multicoloured prisms flashed from it. Sorciero held out his hand and Mark gave him the jewel.

'Probably came out of a Christmas cracker,' Mark said.

Sorciero shook his head, and turned the ring around and around. 'It looks real to me Mark; there's a mark in the metal. It's 18 carat gold. My wife taught me that. She likes nice jewellery. I'll take it to Tequila with me and have young Chang look at it. Here, in the meantime, take good care of it. Your 'bonne chance' certainly seems to be consistent.'

'Oh Mun, you must come here more often. First the gold teeth, the pouch with the coins and now this ring! I wonder what will happen next.'

'We'll eat the fish if it's ready, that's what's next,' Sorciero said and settled himself on the folding chair. 'It smells good to me.'

'Patience, Mun, just a few more minutes.' Mark turned the grouper and then took out some plates and silverware from the cool-box. He opened a couple of

Mangoreen juice bottles and some water. Sorciero watched the young man take the fish off the grill and fillet it.

'Eat, Mun,' Mark said and started on his own portion.

It had been another fruitless day in the search for oil. Sorciero stood under the shower and hummed a lullaby he usually sang for his son. He missed his family and was tempted to abandon the project and return home. He felt sorry for Etienne Bonheur and the islanders. They were pretty poor, and there was no prospect of prosperity in the future, unless something dramatic happened. It was rather like it had been in Tequila, before the earthquake and the miraculous advent of Petrolia. He was still troubled by the effect the Mangorenian women had on him, and he seemed to be constantly plagued by an erection. Wherever he looked, there were islanders unashamedly smooching around.

He dried himself and put on a clean shirt and cotton trousers. His dread locks were neatly caught in a band behind his neck, and he had shaved and neatly trimmed his moustache. A gong announced that drinks were about to be served in the drawing-room of the presidential Great House. He slipped on his sneakers and hurried down the grand staircase. There were a few guests assembled again for dinner, and Sorciero was introduced to Mrs. Bonheur's sister, Josephine. She was a tall, slim woman with intricately plaited hair, and shining red lips and nails. She was wearing a clinging silk dress, which left little to the imagination. She batted her purple shaded eyelids at Sorciero and said:

'I've heard so much about you, Sam,' she murmured.

Sorciero cleared his throat, and tried to extricate his hand from her grasp. He nodded and smiled, but thought that he would die of embarrassment as his dick lifted the light material of his trousers.

'Let us sit down, Ma'am, I've had an exhausting day inspecting your lovely island,' Sorciero said and hurriedly sat down. 'Now tell me all about yourself. Do you always live here, on the island, or are you visiting?'

'You must call me Josephine, and, I'll tell you all about myself, but first let me get you a drink. What would you like? A rum punch, or a shot of Tequila to remind you of home?'

'I think I'll just stick to water tonight, thank you.'

Josephine beckoned to the butler who was carrying a tray of filled glasses over, and Sorciero lifted off a tumbler of water.

'I see that you like our water, Sam?'

'I don't know whether I do or don't. I think it grows on one.'

'That's exactly what I tell my brother-in-law. It's an acquired taste.' She squinted over at the President and lowered her voice. 'Of course, it can't really compare with Perrier or San Pellegrino. He doesn't like to hear that though, so hush,' she said conspiratorially. Sorciero nodded agitatedly, and gulped his water. 'Would you take me along on one of your drives? I would love to see how you work,' she continued. Sorciero choked on the water and coughed. Etienne Bonheur walked rapidly towards him and clapped him on the back.

'You alright, Sam?' Sorciero nodded and wiped the tears from his eyes. 'Josephine often makes men cry, don't you, my dear,' Etienne smiled genially at his sister-in-law. 'I suppose she told you that our water is not up to Perrier standard, but she's wrong.'

'Oh, Etienne, you have long ears indeed'

'No, my dear, I can lip-read,' the President

answered. Josephine pouted prettily and looked chastised.

'Etienne, I've asked Sam to take me on one of expeditions; I've never seen a diviner at work. You don't mind?'

'That's entirely up to Sam; but I warn you Sam, although you are a powerful Sorciero, she might bewitch you before you even know it. She has powers of her own, our Josephine.'

The butler announced dinner and Sorciero led in Josephine. She leant lightly on his arm and her expensive perfume wafted to him. He silently started an incantation, praying that this cursed organ of his would lie down and stay down. Over dinner she informed him that she was a great, great, granddaughter of the Empress Josephine, and that she had an Hotel Particulier in Paris, near the Invalides, where she spent a good deal of time. Sorciero had no idea who the Empress Josephine had been, nor why her great, great granddaughter should have a hotel, where invalids were. It sounded more like a hospital to him, but he had learned to keep his mouth shut when conversation got above his head. He smiled at her, and told her that he had never been to Paris, but that he found New York very fine. She agreed, and he told her about the United Nations meetings he had attended. She seemed suitably impressed, and toasted him with her champagne. By the time the dinner was over, he had agreed to take her along the next day, and arranged to meet her in the hall at 8.30 am. He excused himself and hurried to his apartment. He ripped off his trousers, flung himself on the bed and finally satisfied the Satan riding in his groin.

Chapter Four

Maria Beatriz tossed restlessly in the emperor-size bed. It seemed alarmingly vast and empty without Sam. She always missed him terribly when they were separated. It hadn't happened very often during their marriage, for which she was thankful. She knew that Sam missed her exactly the same way and that he had never been unfaithful to her. This time however, she had a distinct feeling of impending treachery. She glanced at the clock and decided that 6 am was too early to rise. She turned over on her side and tried to calm her spirits. After all, what woman could be so enticing on that godforsaken island of Mangorenia, she told herself over and over again. She reared up suddenly, and threw off the light blanket. Any woman might do it, she thought; Sam was so guileless and inexperienced in the ways of the world. She considered herself far more worldly wise than her husband. She looked at the clock again and decided not to wait any longer. Old Crone was an early riser and she had to be consulted. Maria Beatriz lifted the receiver and dialled the number.

'Old Crone, good morning. Are you awake?'

'Yes Ma'am. Is every thing alright?'

'I don't know; I have such strange feelings, bad feelings.'

'That's because Sorciero is away. I miss him too and so do the clients,' Old Crone tried to comfort her.

'It's more than that, Old Crone.'

'Did you speak to him on the phone, Ma'am?'

'Yes, and he seemed quite happy, but I think something odd is going on there. I think we should go there and find out.'

'Men don't like to be followed around, Ma'am. Sometimes, it's better not to know everything, so Confucius says'

'Please, don't give advice with your Chinese proverbs at 6.30 in the morning, Old Crone. I want to go there and you're coming with me. When can you be ready? I'll send a car for you' She heard Old Crone sigh.

'I am ready now, Ma'am.'

Old Crone Chang gingerly climbed up the steps into the helicopter that was parked on the helipad of the presidential villa. Maria Beatriz was already sitting in the co-pilots place, her earphones in place. Old Crone squeezed herself into the seat behind, and put on her seat-belt. She was not fond of helicopter rides and steeled herself for takeoff. She was thankful that it was only a 15 minute flight. The helicopter took to the air and headed south. There was nothing but ocean between the two islands, and the colours were breathtakingly vivid and beautiful. It was deep ultramarine which turned into bright turquoise, as they neared the shores of Mangorenia. The customs and immigration official, revolver at the ready, was standing on the tarmac as the helicopter landed on the tiny airstrip. He recognized the emblem of Tequila, the crowned agave, which was emblazoned on the side of the craft. He knew then, that it was someone important who had come to visit, and he holstered his gun again. Maria Beatriz jumped down the steps nimbly, and then helped Old Crone down. Maria Beatriz, as usual, was draped in a bright coloured pareo knotted at the waist; a silk tank top and a matching turban swathed around her head completed her outfit. An oversize pair of designer sunglasses covered most of her upper face. Old Crone was dressed in the Chinese traditional black trousers and tunic. The customs and immigration officer looked dumbfounded at the two visitors. He had never seen a Chinese person before, nor a lady dressed in that fashion.

'I am Marie Beatriz Del-Rey, President of Tequila, and this is my friend Mrs. Chang. Take us to the Presidential Great House.'

'Yes Ma'am, but I got no vehicle, Ma'am.'

'Well then do something about getting one, Mun.'

'I could radio up to the Great House to send someone down.'

Maria Beatriz smiled at the officer. 'That's a good idea, Mun. Get us a couple of chairs in the shade while we wait.'

It was Mark who came to collect Maria Beatriz and Old Crone in the Jeep. He had come in early for his appointment with Sorciero and was quietly smoking a cigarette, when he was dispatched to the airstrip to pick up two ladies from Tequila. Just as they were pulling up to the Great House, Sorciero and Josephine stepped out of the front door. Josephine was wearing mini shorts and a halter held up her ample breasts. Her legs were smooth and shapely and her skin a warm bronze. A man's Panama hat was perched coquettishly on her tresses, and she smiled at Sorciero as she took his arm to descend the steps to the gravel path. Maria Beatriz cursed under her breath.

'I knew it; I knew there was a woman mixed up in this somehow.' She caught Old Crone's hand. 'What do you say to that, Old Crone?'

'Nothing, my dear. She is just another guest of the

President and his wife, I suppose.'

'Driver, who is that slut with Sorciero?' she asked Mark. He looked shocked.

'That is Madame Josephine, sister-in-law to the President. Madame Josephine is to accompany Sorciero and me on our tour today.'

'We'll soon see about that,' Maria Beatriz said and jumped out of the Jeep. 'Sam,' she called. 'Over here, Darlin.'

Sorciero's head swivelled towards the sound of his wife's voice. He disengaged his arm from Josephine's hand, and hurried down the steps.

'Sweetness, is it really you?' He swept Maria Beatriz off her feet and swung her around. 'What a lovely surprise. You don't know how I've missed you. Why didn't you let me know? Is anything wrong? The child?'

'No, no, nothing like that. You know me,' she smirked. 'Spur of the moment; that's the way I like doing things.'

'And you've brought Old Crone... something wrong at our consulting room, Old Crone?' Old Crone shook her head and smiled.

'Relax, Sorciero, it's just a visit. We were wondering how you have got on.' Old Crone squinted sideways at Josephine who had joined them. 'Won't you introduce us to your ah... friend, Sorciero?'

'Of course,' Sorciero turned to Josephine, whose smile had become fixed on her face. 'Can I introduce my wife, Maria Beatriz, and my partner Old Crone Chang. We will have to postpone our outing, Madame Josephine. I would like to take my wife and Mrs. Chang in for breakfast.'

Josephine continued to smile graciously. 'Charmed, ladies. My name is Josephine Beauharnais the 7th. Naturally, I forgive you, Sam. No doubt you have a lot to talk about. See you at dinner, then.' She waved her hand languidly and walked down the path into the formal garden.

'She's quite a dish,' Maria Beatriz muttered.

'I suppose so, but so are you,' Sorciero answered and grinned wickedly. 'C'mon Sweetness, up to my apartment, I want to show you something.'

'What about Old Crone?'

'You won't mind waiting a little while, Old Crone. Breakfast is in the dining hall through there. They'll bring you some boiling water for your tea, which you have brought with you, as usual.'

'Will you still go on your prospecting tour this morning? I would like to come along, Sorciero.'

'Sure ting, Old Crone; see yah.'

No sooner were they in Sorciero's suite, than he started to undress and mutter. 'C'mon, Darlin, get these clothes off, I want to make love to you.'

Maria Beatriz giggled. 'What's with you, Sam? You want to show me your etchings? Has that Josephine excited you so much?' Slowly she peeled off her pareo and tank-top.

'Woeh, you sure have a hard-on, Darlin. I think your cock has grown in the last few days, I swear to God.'

'This ain't a laughing matter, Sweetness; I have a hard on most of the time. Look at my balls, they big and hard and swollen too.'

'Just like a bull in the Arena,' she said and touched him. He roared and threw her on the bed. They rocked back and forth until she whimpered, and still he pressed on, unabated, throbbing, remorseless.

'My God, Sam, is this some new magic, or what,' Maria Beatriz panted.

'I don't know, Sweetness, maybe so,' he gasped and finally came in a great burst of fluid.

Half an hour later, they joined Old Crone in the hall.

She was sitting in the overstuffed, oversized armchair, a diminutive figure holding a cup of steaming tea. Her basket, which went everywhere with her, was standing on the floor. She sipped the tea and made a face.

'What's the matter, Old Crone? You look disgusted,' Maria Beatriz said.

'The tea tastes strange. I used my own, but it tastes strange.'

'It's probably the water; you know that they use mostly desalinated water. Thank God that we have never lacked water in Tequila,' Maria Beatriz said. 'Are you ready to go, Old Crone?'

'I have been waiting for over half an hour, Ma'am,' Old Crone answered a little reprovingly and put her cup down.

They left the Great House and Mark drove the jeep to the north. He negotiated the narrow main street of the capital, St. Francois, avoiding some roving chickens, and a goat or two. It was market day, and stands lined the street. The produce appeared to be somewhat limited to sweet potatoes, onions, cabbage, the locally grown spinach, corn, eggs, and of course, Mangoreens. There were some lime and papaya on display, but it was obvious that the container ship had not arrived yet this month. There were also some arts and crafts stands, which displayed raffia baskets of all sizes, and carved animals probably imported from China; brightly coloured hats, some bolts of cloth, a few T-shirts. Cheap pots and pans completed the assortment. There was, of course the water stand. In this hot, dry climate, it was imperative that enough liquids were taken. Women, holding their young children, were ambling along, stopping here and there to make a small purchase.

'You certainly drink a lot of bottled water,' Sorciero commented.

'That's because the desalination water tastes so bad. The bottled water is free, provided by the government. The spring gushes plenty enough for us all,' Mark said.

'Correct me if I'm wrong, Sorciero, but are most of these women really pregnant?' Old Crone asked.

'Yes, they are. It's the population explosion that Etienne was telling us about. Isn't it wonderful? They are all so attractive.'

'C'mon, Sam, since when do you find a pregnant girl attractive? I know better than any one how you feel about that,' Maria Beatriz said and looked at her husband in surprise.

'It's different over here,' Sorciero said simply. He bent towards his wife's ear and whispered, 'they make me feel horny, Sweetness.'

Maria Beatriz suppressed a giggle and lightly slapped his face. Old Crone sat impassively next to Mark and observed the street scene.

'Even all the bitches seem to be pregnant,' she said. 'I can't understand what brings this on.

'We jus love to make love, Old Lady,' Mark answered this rhetorical question happily and honked his horn to get the goats out of the way.

'People call me Old Crone. What about birth control? Never heard of that in these parts?'

'It's like this, Old Crone. We never used to need it before the earthquake. Folks didn't used to fuck all the time, as they do now.' He turned his head back towards Sorciero, and grinned knowingly.

'Where exactly do you want to go today, Boss?' he asked.

'If you don't mind my interfering, Sorciero, I would like to go and look at the spring,' Old Crone murmured.

'That's a fine idea,' Mark said. 'There's a lovely natural pool there, where we can refresh ourselves.'

'But we have no swim suits with us,' Maria Beatriz objected.

'That's OK Ma'am, nobody wears them up there no how.'

'How far exactly is this place? Can you show it to me on a map?' Sorciero asked.

'It's only about fifteen miles on the other side, about a third of the way up the 'Crache Feu,' our volcano. We name the place Oasis, because of the greenery there. It ain't on any map yet.'

Mark drove the jeep out of the town and drove along the coastal road. The beaches were magnificent. Finest white sand spread to the water which was limpid and clear. There was a low crest of foam, where the open sea spilled over the reef. A small plantation of coconut palms put the finishing touch to the perfect picture postcard effect. As they rounded the bluff to the north side, the sand became greyish black, glistening like powdered mother'o'pearl. Mark explained that it was volcanic sand, as opposed to coral and shell on the south side. The road started to climb a bit, and came to a stop at a viewing point. From then on it was only a dirt track and the jeep laboured towards the Oasis, as it was so aptly named. It was an expanse of greenery, like no other on the island. It covered several acres of the slope. A small white shack was in the centre and the road ended there. A van was just being loaded with crates of green bottles.

'Looks just like Perrier bottles,' Marie Beatriz said.

'That's because they are old Perrier bottles. We remove the labels, wash'em out and fill them again. We get a refund of 5 cents per bottle, and of course citizens get the water for free. Tourists have to pay for it.'

'Didn't think you had much tourism here,' Sorciero commented.

'We have the dive motel and the 2 dive-boats which are in the harbour. Then there's the Dock cafe' and bar; you should go and have a drink there. It's quite famous with the yachting crowd,' Mark said proudly. He jumped out of the car and held open the door. 'You must get out here and look at the spring. It's really very pretty.'

They got out of the jeep and walked through an avenue of acacia trees. All around them was tropical vegetation; Hibiscus, Mango, Papaya, Bananas, and a Bamboo forest. They saw the waterfall and heard it splash into the pool. They reached the edge of the water, and caught their reflection in it.

'So that's what we've been drinking,' Sorciero said and took off his sneakers. He waded in, picking his way over the pebbles.

'It's fresh and clean. I'm going in for a dip. Any one coming to join me?' He took off his shirt and shorts, threw them to the edge, and plunged into the pool. He came up gasping for air.

'This is like bathing in Champagne,' he shouted and splashed around like a child. 'Maria Beatriz, you must feel this,' he said and held out his hand.

'I can't come in. No swim suit,' she demurred.

'Nobody wears costumes here, Ma'am, jes strip and go,' Mark said earnestly.

'No wonder that you have a population explosion, if you all go about without any clothes on,' Old Crone chipped in. She removed one of her black Chinese slippers and dipped her tiny foot into the water. 'It is a strange but pleasurable sensation,' she said. Maria Beatriz gingerly waded in up to her knees as well, holding up her pareo. The cool bubbly water gently seemed to massage her legs.

'Turn around Mark and don't peek,' she said, untied her pareo and took off her tank top. She handed the clothes to a disapproving Old Crone, and slid into the water. She and Sorciero bobbed up and down, holding hands.

'This is amazing, Old Crone. You've never felt anything like it,' she called out. 'My skin is prickly all over, and well, even in the strangest places.'

'Come out now, Maria Beatriz, you can't behave like that in public.' Old Crone admonished her sternly.

'No worries, Old Crone, people fuck here all the time,' Mark said, and started walking away. 'Use the towels I've brought down for you and come on to the bottling house when you're ready.' He held out his hand and gently took Old Crone's arm. 'You'd best come along with me, Ma'am.'

The small white building had three ceiling fans which were blowing at maximum strength. Copper pipes came out through the wall and connected with an antiquated bottling machine, which crown corked the bottles when they were filled with water. Two men were standing by, packing the bottles into plastic crates, a dozen at a time. The guests watched the procedure, and expressed their interest and admiration.

'What I would like to know is,' Old Crone finally said, 'has this water ever been tested by a laboratory, as to its purity and contents? It looks pure and clean, but heaven knows what you are drinking there. It could be very dangerous.' She sniffed and wrinkled her nose. 'I hope we have not been poisoned by it. My tea definitely tasted odd. What about the bottles? Are they properly sterilized?'

'Who are you? The inspector for health and hygiene?' Mark sounded a touch irritated. 'We've been drinking this water for five years, and are still going strong.'

'No cholera, or typhoid outbreaks?' Old Crone insisted. 'No dysentery?'

'No Ma'am, nothing like that at all.' Mark picked up a bottle, poured some into a clean paper cup and handed it to Old Crone. 'You must taste it, and tell me what you think of it. Honestly.'

Old Crone made a grimace but complied by taking a small sip. She let the liquid run around in her mouth and then went outside and spat it out. She looked at the undergrowth which stretched right into the bushes and trees. It was like a carpet of dark green moss with numerous minute purple flowers on it. The odour was pungent and reminded her of the meadows of her childhood in China. She vaguely recalled that the plant was used for medicinal purposes, but couldn't put a name to it. She took another sip of the water and swallowed it this time. It was pleasantly carbonated and refreshingly cool, but she couldn't place the strange flavour it left on her tongue.

Mark took out the deck chairs and set them out under the trees. The ladies were invited to sit down and rest in the shade until it was time for the picnic lunch. Sorciero took his divining rods out, and went back along the path towards the pool. Mark walked beside him carrying a note book, a map and pencil. Sorciero had marked down the areas which had already been prospected and written comments about the day's work. Naturally he had left out the discovery of the grave and its contents, as well as the contents of the grouper's stomach. He believed in finders keepers, in this instance. He moved along the path, through the bamboo forest, holding his rod firmly in his strong hands. The rod jerked again and again, but Sorciero knew that it didn't indicate oil. It was water it recognized. When they left the Oasis, the rod calmed down and they proceeded to climb the gentle slope upwards to the rim of the Volcano. The earth was dark and arid. It could have been fruitful, if only rain would have fallen on it. Now the divining rod was still, and didn't register the slightest tremor.

An hour later they were eating their picnic. Mark was disappointed in the day's work. No buried treasure had been dug up; no fish had jumped out of the sea with a jewel in its entrails. Not to mention the nonexistent oil. There wasn't much left to explore on Mangorenia, except the west shore, which was so rocky and dangerous, that it was more like a mountain climbing expedition than a pleasant walk. The only building on the west shore was an ancient lighthouse, which warned sailors of the perils of approaching Mangorenia from that side. It was still in working order and a lighthouse keeper lived there, tending to it.

'So there's nothing up that side, Sam?' Maria Beatriz asked.

'Nothing that I can detect, Sweetness,' Sorciero said, 'unless we go right into the volcano and continue there. What do you say, Mark?'

Mark crossed himself and closed his eyes. 'Never inside 'Crache Feu'; it's 'malehance' to go into the crater. It's dangerous.'

'What do you mean?' Old Crone asked.

'I don't really know; we just don't do it, Ma'am.'

'When did the volcano last erupt?'

'Oh, a long time ago. Before I was born. There are rumbles, sometimes; that means that the Gods are angry.'

'What Gods? Aren't you a Catholic, Mun?'

Mark nodded vehemently, and crossed himself again. 'Sure I'm a Catholic, but there are things the elders sometimes speak of. You should know, Sorciero, the voodoo stuff.'

'Is there something like this going on here, Mark?'

'It is said that now and then there are meetings inside "Crache Feu",' Mark's words became a whisper. 'There is a woman, la Mère Magot; they say her family came from Haiti, long, long ago. She does voodoo, it is said.'

'Who is saying this?' Maria Beatriz wanted to know.

Mark shrugged his shoulders and looked down his nose. He remained silent and chewed on his ham sandwich.

'Well, young man, has the cat got your tongue?' Old Crone chirped in and looked at him expectantly. Mark rolled his eyes and swallowed convulsively.

'It's 'malchance' to talk about it, Ma'am. I should never have mentioned it.'

'Nonsense, young man; we are professionals here. Don't you know that Sorciero is the most famous witch-doctor in the Caribbean? You can tell him anything he wants to know, young man.'

Mark looked dubious and remained stubbornly silent. 'Have you ever been to one of these meetings, Mun?' Sorciero asked. 'I sure would love to participate in a voodoo ceremony.'

Mark shuddered and crossed himself again. 'I don't know nothing Sorciero, it's just rumours,' he said. 'But I can show you where the Mère Magot lives. It's almost on our way home. That's as much as I can do.'

Mark took a detour to get back to the town. The dirt road was potholed and the jeep rattled ominously. They came by a tiny village, which consisted of about five cottages. A woman was digging the front patch of her garden. She had a brightly coloured bandanna around her head. Her ebony skin was pulled tightly over her bones. She lifted her eyes to them as they drove past slowly, leaving a trail of dust behind them.

Mark kept his eyes on the road, and murmured, 'That's her, Mun. Did you see the look she gave us? I hope she hasn't put a curse on us.' Sorciero turned to see whether the woman was still looking at them.

'Stop the car, Mark, and reverse. She is waving at us to come back.'

'Please, Mun, don't make me do that.'

'Yes, you must. You don't want to upset the lady, do you?'

Mark muttered a curse under his breath and reversed. The woman dropped her spade and came to the rickety garden gate. She beckoned with her bony finger, and Sorciero got out and went to her.

'I know who you are; you are the Shaman of the lucky island,' She put her hand on his head. 'You are a good man and your Magic is strong. Why do you want to see me?'

'Just to learn more of your magic, Mère Magot.'

'So, you know my name... I suppose the young driver has gossiped about me.'

'Not really, not gossiped. Talked with respect, rather.'

'Now that you know where to find me, I expect a visit from you, Shaman.'

The drive back through the town went quickly. The market was over and the stands had been dismantled. The passengers in the jeep were strangely silent, and Mark had turned on the car radio full blast. They arrived at the presidential Great House in time for tea. Etienne and Marthe Bonheur were in the drawing-room, and invited them to partake of the smoked salmon canapes and fairy cakes. Etienne Bonheur was keen to hear if any progress had been made during the day's expedition. He seemed disappointed to hear that there was no good news on that score. Excellent host as he was, he invited Maria Beatriz to stay the night and extended the invitation to Old Crone as well. Sorciero was all for it and tried to persuade them to remain in Mangorenia.

'Why don't you spend the night here, Sweetness, and fly back tomorrow. You too Old Crone; after all, Nanny can take care of Junior, but I need you to take care of me here,' he cajoled. Maria Beatriz was about to demur, when Josephine Beauharnais, came into the drawing-room. She was already dressed for dinner in a gold lame number, slit to above the knees on the sides. Her perfect bronze legs seemed endless, and her plunging neckline drew the attention of all present. She embraced her sister, and said that she was invited to cocktails but would be back in time for dinner. It didn't take Maria Beatriz more than two seconds to come to a decision.

'I would love to stay, such fun... You too, Old Crone. Of course I must telephone home and tell them. We must also get a bed for the pilot.'

'I will arrange everything, my dear,'

'I have nothing to change into for dinner,'

'We are very informal tonight. Marthe will give you something suitable. It's only family, so don't worry about that.'

The dinner gong sounded at eight pm, and the guests gathered down in the drawing-room. Maria Beatriz and Old Crone had been provided with all the necessary toiletries, and Maria Beatriz had borrowed an elegant, gold embroidered kaftan from her hostess. They all looked refreshed and cheerful as they sipped their cool drinks. Josephine Beauharnais was back from the cocktail party, and entertained them with an account of the guests there. She had a sharp tongue. Maria Beatriz covertly watched Sorciero and knew that her decision to stay had been the right one. Josephine's flirtatious nature combined with the climate was well-nigh irresistible for any man.

Old Crone delicately took little bites of the lobster thermidor and when the conversation had subsided somewhat, she decided to speak.

'Mr. President, may I ask a question?'

'Of course, Madame Chang.'

'We were very impressed with our excursion today. Our young man took us up to the spring, and showed us the bottling plant.' Etienne Bonheur was chaffed at the praise. 'My question is: have you ever had the spring analysed for purity and whether it is really fit for human consumption?'

'My dear lady, there really was no need to do that. As you can see, we have survived pretty well. The water is fine, and nobody has any complaints about it.' Etienne Bonheur smiled benignly around the table. 'Did you try and dip into the little lake? Marthe and I often go up there and take a dip. We find it very beneficial, don't we, my dear, especially during the great heat waves of the summer months.'

'Sam and I had a dip, Etienne, it was great fun,' Maria Beatriz smirked. 'Do you mind if I stay here until Sam has finished the job? Then we can go back together. Madame Chang can fly home tomorrow; she has her clients to look after.'

'Yes, yes, of course, that seems like a very good plan to me,' Old Crone nodded and continued to eat her lobster. She was ready to go back to the quiet graciousness of her house. She missed her home and seldom left it overnight, except in an emergency. She completely understood Maria Beatriz' motivation however, and approved it fully. Even a great witch-doctor like Sorciero, could be caught in the snares of such a one as Josephine Beauharnais, as she was pleased to call herself. Old Crone Chang excused herself promptly as soon as dinner was over. She took leave of her hosts, and retired to her room, where she made herself a cup of her special tea. An electric kettle had been provided and the ubiquitous bottle of Mangorenia water stood by it. She was getting accustomed to its taste and sipped the tea. She felt tired and decided that she would pack her overnight bag in the morning. Bed seemed a very attractive proposition to her, and after brushing her teeth and plaiting her hair she slid between the sheets and closed her eyes.

The flight back to Tequila was uneventful and Old Crone descended from the helicopter with a thankful heart. It had been an unsettling trip and she was glad to be on familiar ground. One of the presidential cars was at the airport to take her home and she sat back in the Limo and tried to gather her thoughts. She had hardly slept the night before, letting her mind dwell in the past; on her difficult childhood in China; the farm on which she grew up, half-starved and abused. The memories were painful, but it was necessary that she recall. Finally, at dawn, she captured the memory that been eluding her. Yes, she thought, there were some urgent matters she would have to attend to as soon as possible.

Chapter Five

The sun rose and threw its first weak rays of pink and golden light over the ocean. The doves had been cooing since first light and the vegetation was wet with the morning dew. James Harvey de la Cruz stretched and flexed his muscular body. It was time to move, and he swung his legs over the side of the bed, and sat up. He reviewed the program for the coming day in his mind's eye. It was bull day, and he was driving out to the bull pens to inspect the animals he would have to confront at the end of the week. Although James was an excellent fighter, he had deep respect for the bulls, and always had a twinge of regret and guilt at the way the brave animals were baited to death. Miranda had promised to go with him, and that they would spend the day together. He laughed out loud with pleasure at the prospect. The only cloud in his heaven was the shadow Muso Medusa threw on James' parade. It was like the buzzing of a mosquito in the night, hunting for its dinner; relentless and hungry and impossible to swat. He looked at the bedside clock and found that he had only a half hour left before Miranda was due to come. Formerly 'Tradewinds' had only been accessible by sea, but now a narrow, gravel road was being built, which snaked through the hills to the property.

It was almost time and James hurriedly poured the coffee out of the thermos, and devoured the delicate little Danish pastries which had been served with it.

He felt ready for the day, as he jogged down the path to the reception area. The place was starting to wake up to the daily routine. Breakfast was being laid out in the dining area, and the girl at reception smiled at him and nodded towards the entrance. Miranda had just pulled in and waved to him. She looked so young and fresh, that his heart did a cartwheel in his chest, as he quickly walked to the car.

'You look so beautiful, Randy. It's just not fair,' he complained.

'Good morning to you as well, James,' she grinned and showed her perfect white teeth. Her recalcitrant hair was tied back with a brightly coloured bandana, and she smelled of lavender soap and baby powder. He threw his backpack on to the back seat, got into the car and buried his face in her neck.

'God, you smell so good,' he murmured.

'I love you too,' she said softly. 'We only have another ten days left, before I go back to Milan and you to Argentina.'

'Don't even think of it. You are going nowhere without me.'

She smiled and started to drive out of the property. She knew that the chances of her family agreeing to this attachment were slim. They were convinced that it was just a vacation romance, which she would put behind her once she was back in Europe. Miranda came to the asphalt road and she put her foot down on the accelerator. There was hardly any traffic at this time of day. The breeze rushed by their heads and they smiled at each other. The farm where the bulls were reared, was in the centre of the island, and soon they were at the perimeter of the ranch. Most of the bulls were out in the field, grazing. There was one in a heavily fenced pen which had been built to resemble a miniature arena. Miranda drew up outside the farmhouse.

'Welcome, Caramba de la Cruz, I am Juan Rolando' the headman greeted the visitors. 'We are honoured by your visit.' He turned to Miranda. 'Good morning, Missy. Come with me, and I'll show you around.'

They followed Juan Rolando down the path to the pen. A couple of boys were standing by the fence admiring the bull. James looked the animal over for a while, and then picked up a cape, opened the gate and slipped in. The bull slowly turned its massive head and they stared at each other. It was an obvious battle of wills. The animal shook its head and flared its nostrils, then turned away again. James made a raucous noise in his throat and agitated the cape. The bull suddenly swung around and stampeded towards him. Miranda's heart missed a beat as she saw the animal charging. James swirled the cape around and performed a perfect Veronica. There was an enthusiastic 'Ole' from the lads and the foreman. The bull looked dazed, but instinctively turned and pawed the ground menacingly. James laughed and turned his back on him. He walked to the gate and came out of the enclosure. The bull charged against the wall of the enclosure, splintering a piece off it. Miranda ran to James and held onto him.

'God, you frightened me to death, James. Don't ever do that again, please.'

'Dearest, it's my job,' he answered. 'You'll have to get used to it.' He turned to the foreman and said:

'He's a good bull, Juan Rolando. One of our best.'

'Oh yes, Caramba, he is fine, and has recovered well from the journey.'

'What else have you got here for the corrida?'

'We have some home-grown ones, reserved for the President, and a couple from Spain. You can see them in the field, over there. A good bunch this year, Caramba. Would you like to take a closer look at them?'

They walked along the heavily fenced area, and

James made the raucous noise again. The animals lifted their head and stared at the intruder. One of them approached and butted his horns against the fence.

'They are in good shape and quite aggressive. You will be satisfied, Caramba.'

'Have all the others arrived? The picadors and toreros?'

'Yes, they are all here. Would you like to meet them? We eat early here, and will have lunch presently; we would be honoured if you and the Missy would join us.'

'What do you say, Randy?' James asked her. She nodded and they went off hand in hand to the farm house. The lads crowded around James, and Miranda sat down and watched them. She understood a little Spanish but couldn't follow the speedy exchanges. One thing was clear. James was definitely their hero. Miranda went into the kitchen and inhaled the pungent smell of a paella. The ranch cook smiled at her and invited Miranda to taste the dish. Shrimp, calamari and small pieces of chicken were steaming with the rice, the onions, red pepper strips, tomatoes and herbs. Miranda accepted the full spoon and savoured the concoction. She nodded to the cook enthusiastically. The cook smiled, showing a row of gleaming gold crowns. After giving the mixture another go with the black pepper mill, he lifted the dish off the stove and carried it into the dining hall. He filled the plates generously and handed them around. He invited Miranda to cut the freshly baked loaf and they all started to eat. There was a moment of silence; nothing was heard but the scraping of the cutlery and the low murmurs of approval. The foreman handed ice cold beer around. James was visibly happy and smiled at Miranda. He was obviously used to eating with the ranch hands and the bullfighting gang, but had qualms about how Miranda would feel about it.

'This is some paella. I don't know when I've eaten anything as yummy as this in ages,' she said and held out her plate for a second helping. The cook grinned again.

'I like a woman with a good appetite,' he said.

They finished their meal, and Miranda thanked the cook and the foreman for their hospitality. There was a chorus of goodbyes from the ranch hands as James and Miranda made for the jeep.

'Well then, see you at the corrida, my friend,' James said and shook the foreman's hand.

They left the ranch and Miranda drove to the town. She had the Club mail to deliver to the central post office. It was lunch time and the streets were almost deserted. The banks, shops and boutiques were closed from one to four o'clock in the afternoon, but the post office remained open as did most government offices. The heat of the day lay like a thick blanket over the town, and James yawned.

'You're tired,' Miranda said softly.

'Not tired, just bored,' James answered. He laughed at the astonishment in her face.

'It's just a joke, dear heart,' he cried as she pummelled him on the chest.

'Don't you dear heart me,' she said and got out of the Jeep. She took the mail and went into the post office. James followed her shamefacedly and tried to hug her.

'Forgive me, my love.'

'No, I won't. I'm enjoying our first lovers quarrel,' she said.

'When can we kiss and make up?' he asked.

'You'll have to give me a token, and you'll have to redeem it with a brave action.

'What action do you propose?'

'You must come and have dinner at "Tradewinds". It is a formal invitation from my parents.'

'I accept the challenge. What do you want?'

'Let me see,' she looked at him over and then pursed

her lips. 'There's nothing much that you can offer, but I'll have your shoes. Take them off.'

'You can't mean it, Randy. Not my shoes. The asphalt is boiling.'

'Alright then, your shorts.'

'But I've no underwear on,' he protested.

'That's too bad,' she said coolly, enjoying his discomfort. 'OK, I'll let you off the hook this time. Here, you carry the mail, and wait in the car. I've got something to pick up at the American Consulate.'

Old Crone Chang sat up in bed, and sipped her cup of steaming tea. She nibbled at a couple of rice crackers while she waited for her daughter-in-law Mei-Ling, to arrive. They had spoken on the phone and Old Crone had asked her to come in, for what they called a 'conference'. They often had these 'conferences', sitting together in the cool and tranquil family room, in Old Crone's bungalow. Today, Old Crone had decided that the conference was so urgent, that Mei-Ling would come in for an early breakfast. Old Crone heard the front door open and knew that Mei-Ling had arrived. She and young Chang were the only ones who had a key to the place. Mei-Ling knocked at the bedroom door and came in. Old Crone always took pleasure in seeing her daughter-in-law. Mei-Ling invariably looked fresh and bright, and brought a welcoming smile to Old Crone's lips.

'Good morning again. Would you like some tea, my dear? There is some on the dressing-table.' Mei-Ling nodded and poured herself a cup. The smell of jasmine rose from the cup. She pulled up a dainty French Bergere chair and sat down by the bed. She drank some of the tea, and looked surprised.

'What's the matter, my dear?'

'I don't wish to offend you, Mother, but it doesn't taste as usual. Is it a new tea you are trying out?'

Old Crone smiled. 'No, not the tea.'

'Then it must be the water. We should report it.'

'You're right, Daughter, it is the water. You've got a good palate, but we needn't report it. You see, I used some of the spring water I brought back from Mangorenia yesterday.'

'Spring water? I thought the island was as good as dry, except for the desalination plant and the odd raindrop.'

'So we all thought, but it appears that after the earthquake a spring was found on the side of the Volcano. Just as we got oil, they got water. We went to see the site, and it is very pretty, full of bamboo and other vegetation. There is a natural pool there which is very pleasant to dip into I am told.' She smiled archly. 'Sorciero and Maria Beatriz decided to have a dip and were enchanted with it.'

Mei-Ling couldn't quite make it out. 'That's nice, but what is it you really wanted to talk about?'

'Actually, I want to talk about the water. I want to have it analysed as to its properties. I hope that it will not have nasty side effects. We were all obliged to drink it, you know. In fact they are bottling it for home consumption in second-hand unsterilized Perrier bottles, using an antique bottling plant. Can you get a sample to a good lab in the US for examination, without anyone knowing from where it's coming? DHL it, and ask for priority. You can do that, I'm sure, with all your connections.'

'I'll do my best, Mother. You know that I've resigned from the service, but Jason will help.'

'Yes, my Daughter, I know. It is good; with the baby on the way, there was nothing else you could do.'

Mei-Ling took the old lady's hand. 'You are feeling alright though, because of the water, I mean. No sickness or upset stomach?'

'Oh no my dear, nothing like that, for now.' Old Crone smiled faintly and lay back in her pillows. 'Just a little bit tired, because I didn't sleep so well last night.'

'You must have something to eat, Mother. I'll make you a little toast with butter and jam, or would you like me to cook up some tasty rice?'

'A little savoury rice would be welcome, my dear. Stay and eat a bowl with me before you go. I appreciate your company.' She looked at her daughter-in-law affectionately and patted her hand. 'But after breakfast you must promise to take the bottle of water and have it analysed.'

It was past midday when Mei-Ling left Old Crone's bungalow and drove to the American consulate. She parked her car behind the Capponcinis' jeep. She smiled at the young man who was waiting in the vehicle, and entered the consulate. She asked the Marine at the desk whether the Consul was busy. The Marine told her that Miss Capponcini was with him, but that she could go in immediately afterwards. She smoothed her blue skirt down and took a seat. The air-conditioning was on full blast, and she threw her blue knitted cotton cardigan over her shoulders. She wondered why her mother-in-law was in such a hurry to have the water analysed. Knowing her as she did, it would have to be something urgent and serious; unless the old lady was beginning to lose her marbles. It was not like her to stay in bed with her cup of tea, and not go to her appointments especially when her partner was off island. Mei-Ling wondered when Sorciero and Maria Beatriz were coming home and what if anything, Sorciero had divined in Mangorenia. The Marine got up and offered her a cup of water from the cooler. She took it and drank, and mentally compared it to the fake Perrier. Certainly the fake Perrier tasted different.

The door to Jason's office opened and Miranda came out, carrying a large envelope. She embraced Mei-Ling, burbled a good afternoon, and hurried out to the jeep. She looked happy, Mei-Ling thought. Miranda and the young bullfighter seemed to be inseparable these days. She knew about the concerns the Capponcinis had about this new boy-friend of Miranda's. She would ask Jason if he had had any feed-back about the boy's background and the character called Muso Medusa who appeared to be following him around. Jason beckoned her in and she sat down in front of his desk.

'What good wind blows you my way today, Mei-Ling?' Jason said and sat down himself. He looked at her over the rim of has glasses and found that she was just as attractive and neatly turned as always. Her sleek black hair was tied back in a pony tail, and her magnolia skin was smooth and without blemish. She smiled at him and his heart skipped a beat, as it usually did when he saw her. It had been a blow to Jason when she had married her distant cousin Young Chang, a few years ago, right after the earthquake. Mei-Ling and Jason had worked closely together then, to bring the Mafia, which had practically invaded the island, to its knees. She was an undercover CIA agent, and reported to him. Now there wasn't much to do on the island, and she had resigned from the agency. Jason had been the US consul for the last five years, and was expecting to be moved on very soon. It was going to be hard to leave the island for good; that much he knew.

'You look deliciously cool and lovely, as usual,' he said.

'Why, thank you, Jason.'

'No morning sickness plaguing you, I can see that. In fact, you look radiant. Pregnancy suits you.'

'It's early days yet, Jason. Wait till you see me pushing a great belly before me.'

'Frankly, I can't envisage that at all.' He frowned and shook his head. 'Besides I might not even be here to see it. I expect to hear about my next assignment any day now.'

'You've got to be here; Young Chang and I expect you to be godfather to the baby.'

'I'll come back for that, wherever I may be posted. So, anything particular bring you here, or is that you just want to pass the time of day?'

Mei-Ling smiled and pulled the Perrier bottle out of her hold-all. She put it on the table and he looked mystified.

'Are you giving me a present of a bottle of mineral water?' He asked.

'No, not really, but you can have a small taste if you wish. It is sort of special.'

'You want to know, honestly? I prefer something not quite as bubbly. So what could be special about it?'

'I don't know, but you're going to help me find out. I need to get this water analysed for its properties. Can you send it to the States by the diplomatic courier? I need to have the answer asap; before the end of the week if possible.'

'It's almost Easter week, Mei-ling, I doubt that it is possible. Besides, why don't you read the print on the label or telephone the company and ask them. Be much cheaper and less trouble.'

'Oh, but this is not just an ordinary bottle of Perrier. The contents come from a spring in Mangorenia.'

Jason looked nonplussed. 'I always understood that there was no water on Mangorenia. That's why Tequila funded their desalination plant.'

'It seems that after the earthquake, their volcano, commonly known as "Crache Feu" sprung a leak. It has mitigated the drinking water problem of Mangorenia to a large extent. My mother-in-law brought this sample back from Mangorenia yesterday. You know she accompanied Maria Beatriz there the other day.'

'So they use old Perrier bottles to bottle the water? Doesn't sound very hygienic to me,' Jason said.

'That's exactly what my mother-in-law thinks.'

'I don't understand though what it's got to do with her.'

'She has some bee in her bonnet about it, and you've heard about her bees, so let's humour her.'

Jason turned the bottle around and then held it up to his eyes and looked through it.

'It seems clear enough,' he said, and shook it. 'It's slightly bubbly, I can see that. Alright, leave it with me and I'll try my best. How about lunch at the Yacht Club? We could just make it before the joint closes. No? OK, Young Chang wouldn't like it, I suppose.'

Mei-Ling laughed. 'You're right, about Young Chang not liking it, but as it happens, I have had lunch and am due to give a diving lesson.'

'I really think you should give diving up at this point. Think of the baby.'

'God, you sound exactly like Young Chang and my mother-in-law. The lesson is in my classroom, on dry land, so stop worrying.' Mei-Ling started for the door. 'By the way, what's the word on young Caramba de la Cruz and this clownish Muso Medusa?'

Jason's mien darkened and he cleared his throat. 'I've just given the envelope to Randy. She doesn't know what's in it, but it's not good.'

Mei-Ling stopped and came back to the desk. 'So, are you going to tell me?' she asked curiously.

'It's hard to believe, but young Caramba is a half-brother of the late lamented Claudio Curria. We checked his birth certificate, and found that he was born to Kate Harvey and Romeo Curria. He is illegitimate, because dear Romeo was already married to someone else when he swept Kate off her feet into his bed. Romeo recognised Caramba as his son and mother and son were relegated to an estate in Argentina which belongs to the Curria clan.'

Mei-Ling gasped and sat down. 'You're damned right it's hard to believe. Is it too much of a coincidence for him to be here just for the corrida? He has a legitimate reason, hasn't he?'

'It is a great front. It's odd though, how he managed to ingratiate himself with the Capponcinis so quickly. He was booked into the Carlton Hotel but immediately transferred to the Club Royale.'

'And Muso Medusa?'

'He's an enforcer of the Curria clan,' Jason said. Mei-Ling shook her head in astonishment.

'So there is reason to suppose that there's foul play in the air.'

'We certainly have to be on our guard. As you say, it might be a coincidence, but it's hardly likely.'

'So what are we going to do?'

'Be on our guard, closely watch both Caramba and Medusa, prevent any other Mafiosi from entering the country. They just might try a little mischief during the Semana Santa, and scare the tourists way from Tequila.'

Mei-Ling looked at her watch, and made for the door again. 'Keep me informed, and if you need any assistance, Young Chang and I are happy to help.'

Miranda ran into the house and called for her father. Her parents were sitting on the deck having their sundowner. Maurizio and Domitilla came to the island every year for two months before Easter, and stayed until after Easter Monday. They loved their house on the cliffs overlooking their bay, and after a busy day at the Club Royale, sunset was the best time of day for them. They had a first class manager at the club, but once a year Maurizio Capponcini travelled to Tequila to take stock of the place. He did this with all the Clubs Royale, he had established, but Tequila remained his favourite. He had spent many months there, waiting and hoping that his financial and marital problems would be resolved. It had been a bitter sweet time, which had culminated in the extraordinary adventure of living through an earthquake and discovering a gushing oil field on Petrolia; Petrolia, the small atoll which had been born that day, pushed to the surface by the violent forces of nature. 'Tradewinds' was the dream house Maurizio and Domitilla had bought, had furnished and decorated with loving care. It had been, Maurizio used to say: like getting married again without a divorce. They sat on the deck of the house, holding hands, like newlyweds, gazing into the sunset.

'Hey, you two, remember me?' Miranda woke them out of their reverie. They smiled at their beautiful daughter, and she thrust the envelope into her father's hands. 'With the compliments of Jason. He said it was important, and you were to look at it right away.'

'Did you have a pleasant day, Randy?'

'It was totally exhilarating, Darlings. You should have seen James eyeballing the bull, then turning his back on the animal and walk away. It scared me half to death. We had lunch with the ranch hands; they all think that James is the coolest thing since John Lennon, at least.'

'It seems you do too, Kitten,' Domitilla said.

Miranda nodded and turned to go into the house. 'I'm going to have a bath and get ready for dinner. James will be here in an hour. You haven't forgotten he's coming to dinner, Mother?'

'Of course not, Kitten, how could we,' Maurizio smiled and started to open the envelope.

'I hope that Cook is preparing something special.'

'Why don't you go and have a look.'

Miranda laughed. 'OK, but she doesn't like it. She always throws me out. Says I bring chaos into her kitchen.' She stretched her arms above her head and cried: 'Life is wonderful; I want this week never to end. Tomorrow we'll take the boat to Petrolia; James is dying to see the place. I've told him all about what happened there during the earthquake. Do you think I can ask for a permit to go on shore?'

'I doubt it, Kitten. You know how strict security is. Get ready, otherwise your guest will be here before you're in the tub.'

Maurizio took out the sheets of paper from the envelope and started to read the documents. Domitilla got up and bent over his shoulder. She looked at the typewritten page. She paled under her tan, and gripped Maurizio's shoulder hard.

'Madonna, I can't believe it; there must be some mistake,' she whispered.

'I'm afraid that it is most probably true.'

'What are we going to do?'

'Randy must be told; poor child, she will be devastated,' Maurizio murmured.

'She won't believe it.'

'It is hard to believe, Domitilla.' Maurizio shook his head. 'I'll bring it out into the open tonight. Let's see what the young man has to say. We'll give him a chance to explain, or deny it.' Maurizio returned the report into the envelope and got up. 'Let's go inside, I want to phone Jason and hear what he thinks of the report.'

James drove the jeep gently up the gravel road to the house on the hill. He was mindful not to kick up any dust; it would ruin the hibiscus and bougainvillea which lined the way to the house. It was dark now, and 'Tradewinds' was festively lit. The low, somewhat sprawling, Spanish style building, reminded him a little of home. He had dressed neatly for the evening, in a white linen shirt and white trousers. He had caught the sun a little and his hair was bleached flaxen and shiny, tied at the back of his neck. He carried a small posy of flowers, and a box of candy. His mother had obviously taught him about the fitness of things.

Miranda saw him get out of the jeep and thought that he was the most beautiful man she had ever seen. She came to the door and opened it before he had time to knock or ring the bell. She smiled at him and his heart melted, as it had done the very first time he had seen her. She also wore white linen. A sheath so simple, that it had had to have been wildly expensive. Her hair was piled high on her head, and a jewelled barette tried to keep it in check. Her eyes were shining, and he inhaled her fresh perfume as he bent towards her to kiss her on the cheek.

'I've brought some flowers for your mother and some candy,' he whispered into her ear, before he gently bit into her lobe.

'That's very civil of you and all that, but what have you brought for me?' she teased.

He beat his chest. 'Only my heart. It's misbehaving terribly. It's just dying to jump out of my breast and join yours. Will you take it, blood and all?'

'And I thought you were being so romantic... Now here I am about to be covered with gore on my pristine white designer dress.' She rubbed her cheek against his. 'Come along, you'd better deliver the flowers before we crush them, and the candies melt in the heat.'

She took the candy from him and they went through the large living-room hand in hand to the deck. Maurizio was standing at the wet bar mixing drinks; Domitilla stood beside him, a worried frown on her forehead. James walked up to her and she held out her hand. He took it and brought it to his lips, the way his mother had taught him. He smiled up at her and gave her the posy.

'Thank you for inviting me to dinner, Madame Capponcini.' He turned to Maurizio. 'Mr. Capponcini, good evening.'

'Good evening, Caramba, or James, as Randy tells us your name really is. What can I give you to drink?'

'Iced tea or lemonade would be great. I usually don't drink alcohol for several days before a fight. Besides, I'm driving and it's your jeep. I have to be doubly careful.'

'I'd better put the chocolates in the refrigerator; James brought those as well,' Miranda burbled.

'That's really kind of you James, but it wasn't necessary,' Domitilla said quietly.

'No trouble, Ma'am,'

'I think I'll just go to the kitchen for a moment. Why don't you sit down with Maurizio, and keep each other company.' Domitilla followed her daughter into the kitchen and absently looked at Cook, who was busy with her pans. Miranda put the chocolate candies in the refrigerator and put her arms around her mother.

'Isn't he the most beautiful man you've ever seen, Mama? Except for Papa, of course. James is so gentlemanly, polite and old-fashioned. Almost as though time had stood still on the ranch in Argentina. Do you realise that he has never been away from South America'? That he's lived all his life on this estate with his mother, was schooled at home by a resident tutor, and practically had no friends except the stable boys. He's only been to Buenos Aires and Caracas for the bullfights.'

'He's certainly had a girl friend or two, Randy. Remember the photos in the "OLA"?'

'Of course he's had mistresses, Mama. It's only natural for a man to sow his wild oats. That's what you've always said.'

'Changing the subject, you talk about his mother, Darling, but has James ever said anything about his father?'

Miranda nodded and picked up a radish rose. She dipped it into the dish of salt which stood near the stove.

'He never knew his father. He died before James was born. How sad...' Miranda chewed thoughtfully on the radish. 'Not to have a father. I don't know what I would do without Papa. I couldn't imagine life without the both of you.' She put her arm into her mother's. 'Let's join them. I'm sure that cook will do very well without us.'

'Do you love him, Kitten? You're far too young to know about love.'

'We love each other, Mama. You must believe me. This is really serious.

'That's bad, Kitten,' Domitilla said sadly. 'That's really bad.'

'Oh, Mama, he wants us to get married. He's going to ask Papa tonight.'

'Why are you rushing things like that, Kitten?'

'You've forgotten what it's like to be in love, Mama. How long did you wait?'

Domitilla smiled. 'Not very long, but Papa was a mature man, who knew what he wanted.'

'I presume that you knew what you wanted too, otherwise I wouldn't be here today. You were hardly older, were you.'

Domitilla tried to smile again, but didn't succeed. 'I don't think Papa is going to be pleased, Randy. Prepare yourself for some hard words.'

Miranda tossed her head rebelliously. 'Hope you're wrong, Mama, because it would break my heart if you didn't consent. Let's go and join them; James is bound to have spoken to Papa by now.'

Maurizio and James were standing on the deck, holding their glasses. James cleared his throat, and looked at Maurizio.

'I would like to marry Miranda, Sir,' James said, 'but before you say anything, there's something that I have to tell you. It's an ugly story about me and my family.'

'Go on, I'm listening.'

'I don't know where to start, because strictly speaking, I don't know the beginning, except for what I've been lead to believe. My mother's maiden name is Harvey. That's what appears on my birth certificate together with Curria. My mother unknowingly fell into a bigamous marriage with my father, who was already married. They were married aboard a ship which was sailing to South America. They set up home on an extensive ranch in the depth of the Argentine pampas. My father had to leave on urgent business and sailed back to the USA. He apparently never returned. I was born on the estate, and lived there in comparative isolation with my mother. He was a generous man, and we never wanted for anything. It was not an unhappy childhood, but my mother was very lonely and never left the estate. I didn't have any formal schooling; my mother taught me the basics at first. Then had a tutor for the more difficult subjects like mathematics and science. My mother started to breed fighting bulls and it was natural for me to take up bullfighting from an early age. It seemed that I had a talent for the sport, and was permitted to perfect the skills needed for it. I was allowed to travel to Buenos Aires and Caracas, always under strict supervision, and fight in the bullring.'

'It does not seem to me, that apart from not having a tangible father, you had to endure great hardship,' Maurizio said slowly. 'As for companionship, you had several young ladies in tow, according to the gossip glossies.'

'True. My minders threw some glossy girls my way. They thought that I needed to become a man in the bedroom, not only in the ring. I never loved any of them.' He dropped his eyes. 'The worst is still to come, Sir. A little while ago, a man came to the ranch and proclaimed that one Romeo Curria was my father. My mother didn't deny it, so I must assume that it's the truth.' He looked up at Maurizio, and waited for his reaction. There was no surprised exclamation from Maurizio. 'Romeo Curria demanded that I should take up some vendetta I knew nothing about, and avenge the death of his eldest son, my half-brother, Claudio Curria, who had died on Tequila during the earthquake,' James continued haltingly. 'He threatened to torture and kill my mother, if I didn't do as he asked. I was the ideal weapon, because I could travel freely and had no obvious connection to the clan. So here I am, not knowing which way to turn. Help me, Sir, they hold my mother prisoner at the ranch until I fulfil my mission.

'Is this the whole truth, James?' Maurizio finally asked.

'Yes, Sir, it is,' James replied simply.

'And you dare to tell me that you love my daughter? By your own admission you came to Tequila to ruin my family, even assassinate us; destroy Petrolia, and take over the island. The whole thing is impossible!'

'It was a bit optimistic, I admit, to think that I could do all that single-handedly. I wasn't willing, I can assure you, to harm anyone; they forced me to agree to it. I had no choice.'

'There always are choices, young man.'

'They threatened to harm my mother, Sir, if I didn't comply. She means everything to me. She reared me on her own and we are very close. I never saw my father. He was a vague figment of my imagination. My mother didn't even have a photograph. I knew that she was scared of him, because he kept men on the farm to check on us.'

'Men like Medusa?'

'Yes, but I had never seen him before he showed up here and dogged me constantly.'

'You seem such an honest and upright young man. I cannot understand how you could ever think of doing it.'

'That's exactly it, Sir. I just thought about it. When it became reality, I knew that I could never do it.' James' distress was palpable, and Maurizio felt a twinge of pity for the young man. 'I don't know how to tell Randy, Sir. I love her so much.'

'You will have to tell her, the sooner the better. My wife and I already were aware of the facts!' Maurizio pointed to the envelope on the coffee table. 'The American Consulate has a file on you. Not everything you have told me is in here, but you did put some flesh on the skeleton. I'm glad that you made a clean breast of things, before we challenged you.' Maurizio glanced towards the kitchen and saw Domitilla and Miranda in the doorway. Miranda ran across to him and threw her arms around his neck.

'Well, Papa, am I an engaged girl?'

'Engaging certainly, but not engaged,' Maurizio tried to defuse the tension. Miranda pouted prettily and turned to James.

'You've probably forgotten to bring the ring, James. Is that why I'm not engaged?'

'I do have the ring, Randy. It's not that.'

'C'mon then, let's show it to Mama and Papa. James bought it at Young Chang's this afternoon. I haven't seen it yet, but it's supposed to be awesome!' Miranda burbled.

'Why don't we sit down for a moment, I'm sure that dinner isn't quite ready yet,' Maurizio said. 'We have to talk.'

They sat down and Miranda held on to James' hand. 'I suppose you're going to tell me that we are too young to get engaged, let alone get married, but we're prepared to wait, aren't we James? Not too long of course. You and Mama certainly didn't believe in long engagements.'

'Miranda, this is serious. We want you to look at this document carefully, before we go any further.'

'Don't tell me that this is one of those prenuptial agreement things,' Miranda said. 'It looks like the stuff I brought back from the consulate for you.' She picked up the sheets. 'Do I really have to read this?' She looked at James and he nodded. She started to read and suddenly let go of James hand. She turned the page and her eyes filled with tears. She looked at James in despair.

'Is this really true, James?' He nodded miserably.

'I never want to see you again,' she sobbed. 'I will never marry you.' She started up and ran into the house. They heard a door bang shut. There was a definite finality about the sound.

'What am I to do?' James asked helplessly.

'I can't solve that problem for you, James. The way I see things is, that you're going to be asked to leave Tequila within the next 24 hours. You won't have to worry about Medusa; as you saw in the document, he is due to be deported on the evening flight to Miami.' Maurizio consulted his watch. 'He's probably already in the air.'

'I'm not worried about Medusa, Sir. It's my mother.'

'You'll go home and take care of her.'

'More likely they'll take care of us, if you see what I mean, Sir. Besides, there's Miranda. If I can't see her here, I'll follow her to Milan or anywhere else she goes. God, what a mess. It's just not fair!'

James looked around wildly, like a cornered animal. He ran into the house shouting for Miranda. He knocked furiously at the bedroom door. It remained closed to his entreaties. He sat down in front of it and hung his head. Maurizio stood beside him.

'You can't stay here all night, James,' Maurizio said.

'There must be something I can do to make up for this disaster!'

'The best thing is to leave it alone, for now. Go back to the Club and we'll speak in the morning.'

James shook his head violently and remained sitting on the floor. 'I'm not going back to the Club. That's where they'll come looking for me, if they want to deport me.' He looked at Maurizio imploringly. 'Can't I stay here?' Maurizio put a hand on his shoulder. 'Sorry, but that's not on, James. You've got to leave now, but you can go and stay on the Cat. It's anchored in the bay. Don't come on deck in the morning. Stay out of sight and I'll bring you some breakfast. That's the best I can do for you now.'

James drove the jeep to the Club. The roads were quiet and there was hardly any traffic. The breeze fanning his face felt good, and he managed to regain his composure. He was tired now, and just wanted to get to a bed and close his eyes. He passed the gates which led to the Club and slowly drove to the car park. He avoided the main entrance to the lobby and slipped through the flowering hedge to the path which led to the sea. He saw a few couples still sitting on the terrace, sipping their coffee and cognac. He took off his clothes and shoes and stayed concealed behind the hibiscus bushes. When the last guests had left the terrace, he crept down to the sea, and swam swiftly to the Cat. On board he towelled himself down and then lay on one of the bunks. He could detect Miranda's scent on the pillow, and knew that he had chosen the right one. He curled himself up into a ball, shut his eyes and conjured her up beside him. He went to sleep dreaming that he was holding her in his arms.

Chapter Six

Sorciero kicked back the sheets and gazed at his erection which had literally lifted the bed clothes to a new high. He reached for Maria Beatriz in the semi-darkness, but couldn't find her. A sliver of light showed through the blinds. He turned onto his side and called out her name. She came in from the bathroom and sat down beside him on the bed. He pulled her down and they snuggled up together. He groaned as he eased into her and thrust mightily until they fell limply apart.

'Do you think we should rent a holiday home here, Lover?' Maria Beatriz purred as she stretched her limbs. 'The climate definitely suits us.'

Sorciero reluctantly got out of bed and ambled to the bathroom. 'Yeah, it must be the climate. Got to hurry now, Sweetness. Last day of divining. Coming with me?'

'Sure thing, Lover. I won't let you out of my sight until we're safely back home.'

'You coming to see Mère Magot with me this afternoon?'

'I suppose it's safe enough to leave you with that old harridan,' Maria Beatriz said. 'On the other hand, I don't trust any of these females here.'

Sorciero laughed and held his hand. 'Give me a rub down in the shower, Sweetness.'

'You know where that will lead us.'

'Mmm... any complaints?'

'Are you goin to go messin with me again, Lover?'

'You bet your sweet life, my Darlin, you bet your sweet life I am.'

After another day of fruitless searching for the liquid gold, Sorciero and Maria Beatriz were on their way back to the presidential Great House. The driver Mark crossed himself and reluctantly stopped at Mère Magot's cottage. His passengers got out and rang the small brass bell which hung on the garden gate. The cottage door opened and a seemingly disembodied hand beckoned them in. Mark rolled his eyes and called after them.

'You sure you wanna go in there, Mun?'

'Don't be such a scaredy cat, Mark,' Maria Beatriz mocked. 'We're going to be fine. Do you want to come in with us?'

'No Ma'am, not me. I'll wait here.'

The cottage was dark inside, except for some burning candles. The scent of incense and herbs emanated from the embers in the brass bowl, which swung from the ceiling. The shutters were tightly closed, blocking out the daylight completely. Mère Magot sat on a tall chair, in front of a kind of butcher's block. There was a small, multicoloured cockerel in a basket by her knees. The Mère Magot was dressed in a black kaftan and her head was swathed in a black turban. Only the whites of her eyeballs were clearly visible. They could see that she was uttering some incantation. Her lips moved and a monotonous muttering sound came from deep within her throat. She started to turn her head round and round, clapping her hands. Suddenly her head fell forward and she seemed to fall into a trance.

'Do you think she's alright, Sam?' Maria Beatriz

whispered in awe. 'You better take a look, Lover.'

'Naw, she just in a trance, I think. Let's just wait and see what's going to happen.'

'We could go a little closer, maybe?' Maria Beatriz said.

They tiptoed towards the woman, who slowly raised her head and stared at them. Her white teeth flashed in a distorted smile.

'Welcome, great Sorciero, we shall make great vodoo together.' She bent down and picked the cockerel out of the basket. It struggled furiously and squawked loudly. She held it out to him. 'The honour is yours, Sorciero'.

'What am I to do with it?'

'Sacrifice it, of course. We need blood for the magic.' She picked up a meat cleaver with the other hand and laid it on the wooden block. Maria Beatriz took a step backwards and pulled Sorciero with her.

'We don't do no animal sacrifices in Tequila, Mère Magot,' he said haltingly.

Mère Magot cackled. 'Go on with you, Sorciero. You never killed a chicken for your mother's pot?'

'Well yes, but I don't chop its head off, I wring its little neck.'

'We need the blood for the real voodoo; blood spurting all over us.'

'You ain't going to do this, Sam, are you?' Maria Beatriz whispered.

Sorciero took a step forward. He gingerly took the terrified fowl from the grinning woman. The little cockerel put up a great fight and pecked furiously at Sorciero's hand.

'That hurts,' he said and let go, nursing his thumb, which had been cut by the cockerel's beak. The cockerel escaped and landed on the beamed ceiling, still squawing loudly.

'Now what are we going to do? No blood, no voodoo.' Mère Magot eyed her visitors grimly. 'You go get him, Sorciero, you big strong mun. There's the ladder.'

Sorciero reluctantly opened up the ladder and climbed a few steps up. He reached for the bird, which instantly took off and landed on the top of the cupboard. Try as he might, Sorciero couldn't lay a finger on the cockerel.

'That's enough, Sam. So we won't see the voodoo. We gotta go,' Maria Beatriz cried, and waved her hands about her head. 'This cursed chicken wants to sit on my head now.' She tried to protect her hair, while the bird circled around her head.

'Ugh... this is too much; now its crapped on my hand!'

'C'est bonne chance, Madame,' Mère Magot said, clapped her hands and made cooing noises. 'Viens, mon petit cocorico, retourne chez ta Mère Magot,' she chanted, and held out her hands. The cockerel calmed down instantly, flapped its wings, and settled in her palms.

'You see, great Sorciero, today he won. We will spare him.' She stroked the cockerel's head, smoothing his ruffled feathers.

'It seems to me that he always wins, doesn't he,' Sorciero said.

'You are right. He is a great and cunning fighter. He has survived two whole years, haven't you, my pretty coquelet.' She sat the cockerel back in the basket and produced a crystal ball from underneath the butcher's block.

'I will try and tell the future. Come closer, Madame. It will concern you.' She gazed intently at the crystal ball and put out her horny hand.

'You and Tequila are in grave danger, Madame.

There is bad stuff hanging in the air. Bad men are on the island, and more will be coming. Look to your security and guard precious Petrolia. Beware the theft of your child!' Maria Beatriz shrugged her shoulders and wiped her hands on a piece of Kleenex.

'You must take this seriously, great Sorciero. Tell your lady that my crystal ball never lies.' She stared in to the ball again and hummed tunelessly. 'I also see that you are going to bring prosperity to Mangorenia, Sorciero. Great prosperity.'

'But I didn't find oil, Mère Magot,' Sorciero protested.

'Whatever... it's clear to see that there is something in the ground.'

'We're going to be touched for a loan again,' Maria Beatriz murmured. 'That's where the great prosperity will come from.'

Mère Magot shot chiding glance at Maria Beatriz. 'Do not mock my predictions, Madame. I am not as famous as the great Sorciero, but my magic is powerful. You, Madame, are with child. You conceived it this morning, and it will be a girl.'

'That's not surprising. Everybody seems to be pregnant in Mangorenia,' Maria Beatriz answered. Mère Magot smiled, showing her perfect dentures. She got up and bowed to her visitors. 'You must leave now, I am tired.'

Soricero put his hand in his pocket and took out his wallet.

Mère Magot shook her head. 'We are colleagues, great Sorciero, a fee is not accepted.'

'That's very generous of you, Mère Magot,' Sorciero said.

'I have a wish though... I want to be present at your Happening. Can you arrange that?'

'I knew that there would be a string attached

somewhere,' Maria Beatriz whispered. Sorciero tweaked her arm and she made a little bow.

'You will be welcome, Mère Magot,' Sorciero said politely. 'Come Sweetness, we must go now.' He marched Maria Beatriz out of the cottage and towards the jeep.

'You're hurting me, Lover.'

'You must hold your tongue in check, Sweetness. This old witch might be a powerful Shaman.'

'I know, I know; she makes me nervous, that's all. I'm really scared, Sam. She's a ridiculous sight, but there's something quite awesome about her. We must tighten security at all entries into Tequila, and our son must be doubly protected. I can't wait to get home now.'

They settled in the jeep and asked Mark to step on it. Mark cast them curious glances and then ventured a question.

'You look all shook up; what happened?'

'Nothing much, just routine. Blood sacrifice and predictions.'

'Blood sacrifice?' The driver crossed himself again. 'They say that they sacrifice humans at the rituals.'

'She asked after you, Mark. She knew you were hiding in the jeep. You'd better watch out. She fancies you,' Sorciero teased him. Mark shuddered visibly and stepped on the gas pedal. They made the presidential Great House in record time.

Etienne and Marthe Bonheur were waiting for their guests in the drawing room of the Great House. Etienne and Sorciero went into the study to finalize their business, while Maria Beatriz exchanged civilities with Marthe. The President of Mangorenia and his wife would come over for the Happening, the corrida and the annual Semana Santa Grand Ball. They would be guests of Maria Beatriz and Sorciero and stay at the pink marble villa. Maria Beatriz always felt in awe of Marthe Bonheur, who had been educated in France and always wore French designer clothes to big events. Marthe always regretted that Mangorenia had chosen independence, instead of remaining French territory, and enjoying all the advantages that being a 'Departement' would have bestowed on them. Martinique had chosen to become a 'Department', indeed it had been promoted to a 'Region.' The citizens of Martinique enjoyed all the social security & pension benefits of mainland France. It also boasted the best hospital facility in the Caribbean. Losing all those benefits was a huge price to pay for so called independence. The island republic of Mangorenia was deeply in debt and its citizens poor as church mice. Their hope of finding oil, like their neighbour Tequila, seemed more remote than ever after Sorciero's visit.

The chopper was waiting at the landing strip, and Sorciero and Maria Beatriz were in the air and back on the tarmac at Tequila International Airport in record time. Their driver and Range Rover were there to meet them on the apron, and they were whisked away to the presidential villa. Charlie was waiting for them in the hall and threw himself into his father's arms.
  'Where you bin, Dad? I miss you lots, and Mama too.'
  'Have you been a good boy, Charlie?'
  The child nodded seriously. 'Ask Ling Ling.' The Chinese Amah standing behind him also nodded.
  'Very good boy, master Charlie.'
  'Then you shall have a little present. The President of Mangorania and his lady send you this story book with lovely pictures.'
  'Is it all about Mangorenia?'

'I don't know, Charlie, but we'll look at it when you're in bed.'

They shared the child's supper and then took him up to bed. They sat on either side of him and Maria Beatriz held his little hand. Sorciero opened the book and started to read about the little sea shell which was washed up on the beach, and what happened when a little boy found it and held it to his ear. Charlie closed his eyes and smiled. He was about to go to sleep, when he suddenly opened his eyes again and said:

'Mama, a lady came to see me. She was very pretty, with long, silky fair hair. Long to here, like a mermaid.' Charlie reached for his knees. 'She said she wanted to take me out for a treat, but Ling Ling said no.'

'Ling Ling was right not to let you go out with strangers.'

'She wasn't a stranger, Mama. She said she knew me.'

'You were probably dreaming, Sweetheart,' Maria Beatriz said and stared at Sorciero. Mère Magot's warning was ringing in her ears. Who was this stranger who wanted to take their son out? How did she get past the guard at the gate without an invitation? Luckily Ling Ling was smart enough not to have let Charlie go.

'Never go anywhere Sweetheart, without Ling Ling or Mama and Papa.'

Charlie nodded earnestly and closed his eyes again. He was asleep within seconds, and they left the nursery on tiptoes. They immediately went in search of Ling Ling, and questioned her about the incident. Ling Ling had been Charlie's amah from the day he was born and treasured him. Old Crone Chang had found Ling Ling in the Chinese community in San

Francisco, and recommended her to Maria Beatriz.

'What's this business about a visitor who wanted to take Charlie out for a treat?' Maria Beatriz asked her.

'I had just left Master Charlie in the garden to get him something to drink. He was playing in the sandbox. Joseph was watering the plants and said he would look after him. I was only away for about three minutes getting the orange juice in the pantry. When I returned, she was there, pushing Master Charlie on the swing.'

'What happened then?'

'She waved to me and then walked away towards the gates. Master Charlie wanted to run after her but I didn't let him. She called to him, asking him to come along, but I held on to him.'

'Where was Joseph?'

'He wasn't there. I looked for him and found him behind the hedge. He had been knocked out.'

'When did all this happen?'

'About 4.30 this afternoon. I would have phoned you but you were almost on your way, Ma'am.'

Maria Beatriz suddenly felt a strange sensation in her belly. She knew instantly that Mère Magot had been right about her being pregnant. There was nothing tangible to go on, except this strange sensation. She smiled happily, but then her thoughts were back with Charlie. It was quite obvious to them now, that someone had tried to abduct Charlie.

'Where is Joseph now?' Sorciero wanted to know.

'He's nursing a big bump on his head. I sent some ice packs up to his cottage.' Sorciero and Maria Beatriz phoned the security guard at the gate, and asked the man to bring in the daily report. They waited in Maria Beatriz office for the man. The man had been a security guard at the presidential villa for over thirty years. He was a trusted professional. He had known the Del-Rey family since they were children. He had watched them grow into teenagers and adults. He had shared a glass of tequila with Big Daddy Del-Rey every day until Big Daddy's death. His name was Baldwin, but he was known as Baldy because of the early loss of his hair.

'Come in Baldy. You've heard about this intruder who came into the property this afternoon?'

Baldy grinned. ''Twas no intruder, Missy, it was Missy Maria Dolores. She looked real pretty.'

Maria Beatriz gasped and held on to Sorciero's hand.

'You shouldn't have let her in, Baldy.'

'But she your kin, Missy. I couldn't keep her out,' Baldy stammered.

'From now on, you let no one in, kin or no kin, without calling back to ask. OK Baldy?'

'OK, Missy. I'm real sorry, but I thought, she being your sister...'

'Baldwin, listen to me. No one, unless we tell you.'

When Baldwin left the room, Maria Beatriz started to walk up down in some agitation. She knew instinctively that the presence on the Island of Maria Dolores, her half-sister, as she was pleased to call herself, was bad news. She never forgot the day of her father's funeral, 5 years ago. The President was lying in state at the cathedral. The scene, when Maria-Dolores threw herself on the coffin, and screamed like a demented banshee, was unforgettable. They had had to pull her away and get her out. She had struggled all the way, yelling abuses at the rest of the family. They finally had managed to get her to the ambulance where they injected her with a sedative. Maria-Dolores, who had been the cuckoo's egg in the Del-Rey nest, was known to be the daughter of a ship's captain who had dumped his mistress, a fiery Spanish dancer, in Tequila. The President, Big Daddy, as he was affectionately called, had unwisely fallen for the charms of this castagnette playing senorita and married her. When she gave birth to a healthy strapping baby girl, christened Maria Dolores, three months prematurely, it was patently clear that she had tried to pass off this infant as Big Daddy's. The Spanish dancer left Big Daddy holding the baby, so to speak, and made off with an American pilot.

As the child grew into a sexy teenager, Big Daddy gave up the pretension of fatherhood, and took her to his bed. She immediately began to rule the household, putting all the family's noses out of joint, lording it at the old man's table, and worming large sums of money out of him for her favours. All the Del-Reys hated her, and when the old President died, and Maria Beatriz was chosen unanimously as the next President, Maria-Dolores was invited to leave the island with her booty, on a one way ticket to New York. She swore to take revenge and invoked the 'mala suerte' on to all the Del-Reys heads, before they marched her to the plane. Now she was back, with God knows what nefarious plans in her baggage. If the Mère Magot's warnings were to be taken seriously, then this unwelcome visit to young Charlie was indeed ominous. They would have to concentrate all their efforts in finding where Maria Dolores had taken cover. She might still have a friend or two on the Island who would put her up. Sorciero picked up the phone and dialled Old Crone's number. She usually retired early, but this was important enough to wake her up. When she answered and heard what Sorciero had to say, she immediately agreed that they should meet in the morning with Mei Ling and Jason at the American Consulate.

Old Crone Chang sat in her rickshaw, holding the ubiquitous basket on her lap. The young man was making good time as he cycled down the street to the Consulate. He liked working for Old Crone. She was light as a feather and a generous employer. Besides it did no harm to be working for Sorciero's partner, and he commanded some respect by that mere fact. He had cycled Old Crone's rickshaw from the time when he was only a boy. Now he was almost a grown man, and very protective of his diminutive boss. He stopped outside the Consulate and helped Old Crone descend. She nimbly climbed the few steps into the hall and was saluted by the Marine on guard. She was immediately shown into to Jason Reed's office. Jason greeted her warmly and offered her a chair and some boiling water. It was a well known fact that Old Crone only drank her own tea, which she brewed in her own little pot and drank out of her delicate china cup. She took a couple of small cups, saucers, and a teapot out of the basket, and carefully prepared the tea. She offered a cup to Jason. It was another well known fact that Old Crone's teas were more than beneficial and made one feel good, although no one quite knew why. She grew all the herbs in her garden, which she tended personally. A pinch of her famous herbal teas was one of the more expensive items which the partnership of Sorciero and Old Crone dispensed at their consultations.

A few minutes later Maria Beatriz and Sorciero were shown in, followed by Mei Ling. The last one to arrive was Chief Pereira, the head of the police. They sat around the conference table, where writing-pads and sharpened pencils were laid out, as well as a slim dossier for each of the participants. Jason opened his file and asked them to read theirs carefully. When they had finished, Jason told them that he had sent the dossier to Maurizio Capponcini, as he and his family were directly involved. Then he asked Maria Beatriz to comment.

'Before we discuss this dossier, I want to tell you that the life of our son Charlie is in danger. We believe, Sam and I, that there is a plot afoot to kidnap him.'

'Whatever gave you that idea?' Mei Ling exclaimed.

'It's la Mère Magot; she warned us yesterday.'

Everyone around the table looked mystified.

'Who on earth is la Mère Magot?' Jason asked.

'You tell them Sam,' Maria Beatriz said and held out her hand to Sorciero, seeking his comfort.

'La Mère Magot is the shaman of Mangorenia. She's known for her voodoo powers. We were asked to call on her. She gave me a machete and invited me to sacrifice her cockerel, but he got away, so she looked into her crystal ball instead. She told us that Charlie's life was in danger.' Sorciero turned to Old Crone. 'Why can't I have a crystal ball, Old Crone? Do you think that that is a magic I can learn?'

'I suppose you could try, Sorciero. You could look in the 'Little Switzerland' shops; they might have something suitable there.' She smiled indulgently at Sorciero. 'But no cockerel sacrifices, please.'

'It might be quite impressive though, and we could make chicken soup afterwards,' Sorciero muttered.

Jason cleared his throat and said, 'Can we change the subject? I cannot face the thought of chicken consomme right now. What other proof do you have apart from the machete wielding Shaman's prediction?'

'I'll tell you. It's Maria Dolores; did you know that Maria Dolores is on the Island, Chief?'

'No, Ma'am I didn't. Nothing was reported to me. She must have come in under a false passport.' He lifted his shoulders helplessly. 'Our young border police wouldn't recognize her.'

'She was at the presidential villa yesterday and found Charlie in the garden, and spoke to him. She ran away when our nurse approached her, but the security guard at the gate knew who she was. He didn't dare stop her, as she was family.'

'What about that Muso Medusa? How did he get into the country?' Jason asked.

'He's not on our list, Sir. I'm real sorry, but what with Semana Santa coming, it is always a hard time to keep track of all the trash comes on the island.'

'We know that there's a link between the young Torero De la Cruz and this Muso Medusa,' Old Crone chipped in. 'What if Maria Dolores has been recruited by the Curria gang? She would just love to take revenge on you, Ma'am, after her ignominious departure from the island.'

Maria Beatriz drew in her breath. 'So the Mère Magot was right,' she said, and Sorciero nodded slowly.

'What are we going to do, Jason?' she asked. 'Throw them off the island?'

'We'd have to find Maria Dolores first. As for De la Cruz, there may be a development there. Would you like to tell us, Mei Ling?' Jason looked at Mei Ling.

'The young man has succeeded in ingratiating himself with the Capponcinis and particularly Miranda. They are so obviously in love. At least he gives a very good impression of being in love. Maurizio called me this morning and told me he had got the dossier, and what were they to do. De la Cruz apparently was going to propose to the girl last night, and confessed that he was mixed up with the Curria clan to the parents, before they could challenge him. I told Maurizio that we were having our meeting and that we would keep him informed.'

Old Crone tapped her pencil rapidly on the polished table. Jason invited her to speak.

'Think of the upheaval on the island if De la Cruz were to miss the bullfight. He's every one's hero; a most attractive young man. I think we should keep him here under close supervision and get as much information out of him as possible.' She looked around the table. 'Do you all agree?'

'Yes, if Chief Pereira thinks he can manage that, but what about Maria Dolores? Where will you look for her?' Maria Beatriz said anxiously.

'Yes, Ma'am, we can do that. I'll get my best man on the job. We'll check all the hotels and guest houses too.' He looked at Maria Beatriz. 'We will post some men in the grounds of the villa. Young Master Charlie will be protected at all times, Ma'am.

Maria Beatriz thanked Chief Pereira and the meeting was adjourned. The Marine showed them all out and saluted smartly as Maria Beatriz and Sorciero came by. Old Crone took Mei-Ling's arm and asked her to accompany her back to her house. She also invited Sorciero to join them. She wanted to know all about the rest of Sorciero's stay on Mangorenia. There was some urgent business she wanted to discuss with him, before they opened their consulting room for the day.

The Yoga class was over and the participants had left the gymnasium. Miranda sat cross-legged on her mat and tried to meditate, but she couldn't concentrate on the teachings of that discipline. Tears gathered at the corners of her eyes and ran down her tanned, flawless cheeks. She had cried herself to sleep the night before, and the idea that she might feel better in the morning, had been a vain hope. She couldn't believe that this was happening to her; that James was a deceitful, dangerous man, who had come to the island to harm her and her parents. Her beautiful, tender James, the scion of a Mafioso clan! It was totally absurd, and yet he had admitted it to them. She had vowed to herself that she would never see him again, yet she was longing to run into his arms and be comforted by him. She resisted the urge to go to his cottage and tried to calm herself. She wiped her eyes and got up. She would leave the gymnasium via the back entrance and get to the car park that way. To stay at the club was to invite James's attention. There was nowhere else for her to go but home to 'Tradewinds,' and lock herself up in her room. Her parents seemed to think that she should give James another chance, let him explain; she would have none of it. She threw her windcheater over her shoulders and marched out, shoulders squared, head held high. She made a dash for the jeep and started it. She almost ran over James who was leaping across the car park, trying to stop her. He waved his arms around madly and mouthed unintelligible words, as he ran after her, covered in the cloud of dust the spinning wheels of the jeep created.

At 'Tradewinds' Miranda fell into her father's arms and sobbed. He smoothed her unruly curls until she quietened down. Domitilla took her hand and led her out to the deck. Miranda paced up and down, muttering to herself. Domitilla gave her a glass of ice cold water, and encouraged her to drink it.

'What happened, Randy? Did you see James?'

'He ran after me in the car park, but I didn't stop. There's nothing left to say, is there.'

'Oh I don't know, Darling,' Maurizio said gently. 'I know that it's been a great shock to all of us, but on the other hand, he seems to be in a difficult position. Why don't you at least listen to his story?'

'I can't believe that you're really saying this, Papa.

This man came to Tequila to murder us all; Mei Ling, Maria Beatriz, and Sorciero; destroy Petrolia, or invade it; turn this lovely island into a Mafia stronghold.' She looked at her father incredulously. 'How can you ask me to listen to his story? Besides he's a born Curria.' She shuddered. 'Blood is thicker than water, Papa. You always said that.'

Maurizio looked helplessly at Domitilla. 'There's no getting around that, of course, but he had never seen his father, or had contact with the family until now.'

'How come Jason asked for James to be investigated? Why should he do that?'

'We decided that it should be done. James came to Tequila with a weapon, and that fellow Medusa was always hovering over him. You said so yourself.' Maurizio held out his hand to her and she clasped it. 'So you see, we thought it for the best, especially since the two of you had become so incredibly close. We felt we had to know more about his background.'

Miranda tossed her head and frowned. Her face crumpled and seemed to lose that lovely freshness of youth. The corners of her mouth turned down and lower lip trembled. Maurizio took her in his arms again and cradled her.

'I don't care what his excuse is. I never want to see or speak to him again,' she whimpered.

'I understand, Darling, but that will be quite a problem on this island. Where ever you'll go, you'll find him; he is a guest at the club. Perhaps you'd like to leave for Milan sooner. We could book you onto a flight tomorrow.'

'No, no,' Miranda protested. 'I want to be with you and Mama. I shall just stay here at "Tradewinds" until all the festivities are over and James has left the Island.'

'If that's what you want, Darling,' Maurizio said

indulgently. 'Mama and I are going to town. Anything you want from the shops?'

'Just some hemlock!'

Maurizio and Domitilla drove into the town in silence, watching the landscape move by. She put her hand over his and squeezed it. She sighed deeply and he turned to look at her.

'She'll be fine, Domi; stop worrying.'

'I can't help it, Darling. She looks so sad and bitter. What a blow it is, to find that the man you love is a... I can't find the right word.'

'I know, it's difficult, because he is not a real villain. There is a sort of innocence about him, which prevents one from getting really cross. He could charm the birds off the trees.'

'Well, I wish to God he hadn't charmed our little bird off our tree,' Domitilla exclaimed.

'She's still in our tree and not likely to fly away as yet.' Maurizio said, and deftly manoeuvred the car into a parking spot. They had arrived at the American Consulate to keep their appointment with Jason Reed. James drew up at the same time in a taxi. They entered the Consulate together. They were ushered into the board room and sat down at the conference table. Jason Reed joined them and started the ball rolling, by asking James to tell his story again. James got up and started pacing the floor. He repeated all he had said to Maurizio the previous night. He voiced his concern for the safety of his mother if he didn't fulfil his so-called duty, and his deep revulsion for the task he had to undertake. What was he to do? What did they suggest he do? He looked helplessly at them. Jason looked at Maurizio, who shook his head, and glanced at Domitilla.

'I don't know why you're all looking at me as if I had

all the answers,' she said tartly. 'I'm sorry for you, James, but it's not up to us to get you out of this jam.'

'Right, Domitilla, but we must find out how many there are on the island. It can't only be Medusa. He's here to keep James in line. There must be others who have slipped through the net. We know that for a fact.'

'Who would that be?'

'There's Maria Dolores for a start. It seems that she's on the island and tried to abduct Charlie from the presidential villa, last night.'

'Maria Dolores? On the island? I thought that she was definitely persona non grata.'

'So she is, but somehow she's managed to get through immigration. Chief Pereira is trying to locate her.'

'Who is Maria Dolores?' James asked.

'She's supposed to be the President's sister, but everyone knows that Big Daddy Del-Rey wasn't really her father. He was duped into marrying a Spanish dancer who castagnetted herself into his affection and produced Maria Dolores supposedly as a premature Del-Rey baby. One day, the Spanish dancer disappeared off the Island and left Big Daddy holding the baby. The baby was by then almost a teenager, and she quickly became Bid Daddy's comfort, in and out of bed. That is Maria Dolores in a nutshell.'

'But why would she want to harm Charlie?'

'To take vengeance on Maria Beatriz and Sorciero, who banished her after Big Daddy's death. She had wormed most of the old man's millions out of him, and had come between him and his legitimate children. They were quick to dispose of their father's mistress.'

'No one told me about that,' James said. 'No one ever mentioned kidnapping or harming the boy.'

'It seems that you were not the only player involved

in their plan of attack,' Jason murmured. He looked ruefully at James. 'I don't know what we're going to do with you, James. Are you going to help us get these people? Are you on our side or theirs?'

James stood in front of Jason, and raised his right hand. 'I swear to you that I'm on your side, and that if we can secure the safety of my mother, I will do all that is possible to help you.'

'I'm afraid that you're in no position to make any demands, young man,' Jason said sternly.

'Come, Jason, surely we can do something about James' mother?' Maurizio said calmly. 'You could get her out of Argentina and hide her somewhere.'

Jason looked at Maurizio in astonishment and shook his head.

'You can't be serious, Maurizio. I can't mobilize the agency for such a scheme. I'm afraid that it is quite out of the question.'

'Please consider how helpful James could be; he'd be our undercover man and tell us all about the plotters and the plot.' Maurizio sounded very persuasive. 'It's understandable of course, that he would like his mother to be safe. Wouldn't you feel the same way?'

'Luckily I'm an orphan and don't have to worry about these things,' Jason said sarcastically. 'Where would we take her anyhow?' He turned to James. 'Has your mother got family in the USA?'

'I don't know; she never talked about them. I'm so sorry that I can't tell you more.' He looked desperately around at the others. 'Supposing,' he said then, 'supposing I go to that fellow Medusa, sort of make friends with him and tell him that now I'm ready to do the Family's bidding. He'll tell me what the plan is, and what I'm supposed to do. At the same time I'll find out who else is on the island, how they got in and how they expect to leave. If I do that, will you fly my mother out of Argentina and bring her here?'

'Here, right into the Lion's den?' Jason asked.

'At least we could be together. Would you look after her, Mrs.Capponcini?' James pleaded.

'I think that's a lot to ask of us, young man,' Domitilla said and turned her head away. She studiously avoided looking at him.

'I suppose she'd be as safe here, on the island, as anywhere, and we could keep an eye on her,' Jason commented. 'What do you say, Maurizio?'

'It's up to Domitilla.'

'And what about Miranda? She'll have to know, if Mrs. Harvey comes to stay with us,' Domitilla voiced her concern. 'Randy is deeply hurt and refuses to see or speak to James. Why don't you have her stay with you, Jason? Here at the consulate. Pass her off as your aunt, or something. You're well guarded; there are always a couple of marines on duty.'

James allowed himself the glimmer of a smile. 'I hardly think an aunt would be the appropriate relative. A cousin, perhaps. My mother is a young woman, and doesn't look her age.' He glanced at them all again. 'I won't let you down, I promise. You'll get all the information that you need, and you won't regret that you trusted me.'

They looked at his eager, open young face, then at each other. Jason pursed his lips and beat a tattoo on the table with his index fingers.

'Wait outside, young man, we will have to discuss this in private,' he said then.

When their conference was over, they called James in and told him what they had decided. They would give him the benefit of the doubt, get his mother out of Argentina and keep her at the consulate until the

Semana Santa was over. James would immediately take up contact with Muso Medusa, and find out who else was going to be there to take over, or destroy Tequila and Petrolia. He would be docile and respectful towards Muso. James would remain at the Club Royale, and communicate with Maurizio at least once a day. He would continue going round to 'Tradewinds', and ostensibly be a welcome guest there. He was to go about his usual routine, visiting the ranch, inspecting the bulls. There was one small flaw in these arrangements, and that was that Miranda refused to see or talk to him. She would have to be persuaded that it was vital she at least took a drive with James once a day, to show whoever was watching them, that all was well with the lovers. If all went according to plan, the bad guys would be arrested, tried and deported. Tequila law didn't favour capital punishment, and to keep prisoners incarcerated was an expense the government could well do without. In fact, the prison was usually empty, except for the odd drunk or two who spent the night there and were kept until they were sober enough to go home.

'But what about us; my mother and me? What's to become of us? Will we also be deported and where to? I don't suppose it will be safe for us to return to the ranch. Will it?' James queried.

'We'll cross that bridge when we come to it, young man. That's all for today. It's time you started your new assignment.'

Chapter Seven

James got into the jeep and drove along the coast to the Carlton Hotel where Muso Medusa was staying. It was now eleven o'clock and the sun was beating down on the tarmac. The young Torero prayed silently that he had done the right thing, and that he and his mother would be safe. He jammed a baseball cap onto his head and put on his sunglasses. The road to the hotel ran along the coast and the sea looked inviting, cool ripples breaking against the palm-fringed sand. He turned into the drive and parked the jeep. Now was the moment of truth. Would Muso Medusa accept his turnabout, or would he be suspicious. The bellboy at the door sprang to attention as he saw James walk up. The bellboy grinned and made the swing doors turn at a dangerous rate.

'Ole!' he cried. 'You haven't forgotten the ticket to the bullfight that you promised me, Caramba.'

'You'll get it, don't worry. Say, have you seen Mr. Medusa around this morning?'

'Yes, Mun, he was tucking into his second breakfast on the terrace.'

'Is he alone?'

The bellboy grinned widely again. 'At the moment, yes. He did have a popsy here though, but she left early.'

'Was she a looker?'

'You might say that. Long blonde hair...' The bellboy had a gleam in his eye, a reflected memory.

'Anyway, what do you want with a sleazebag like that Medusa?'

'Watch your tongue, Medusa is a friend of mine,' James said sternly.

'It certainly didn't look like you were a friend of his when you were here last time, Caramba.'

'Nonsense, you're mistaken. It wasn't like that at all.'

The bellboy shrugged his shoulders. Never argue with a client was the maxim of the Hotel. James went through the swing doors, and was hit by a blast of vicious air- conditioning. He took off his cap and went to the terrace. Several guests turned to look at the new arrival. The sight of James Caramba De la Cruz was not to be missed. He saw Muso Medusa somberly drinking his buck's fizz.

'Can I join you, Muso?' James asked politely.

'To what do I owe this unannounced visit?' Muso growled.

'We have to talk, Muso. It's time you told me what the plan is. All I know is that I have a gun, but not what I'm supposed to do with it.'

'Can it be that you've come to your senses at last?' Muso asked sarcastically.

'I've always been prepared to do my duty, Muso,' James said earnestly.

'You could have fooled me, after the reception you gave me.'

'What did you expect me to do? Kiss you on both cheeks? I had to pretend I didn't know a character like you, otherwise I would never have been able to gain the Capponcini's confidence. The truth is that I had never seen you before, and had to be careful.'

'Even when we were on our own?'

'Walls notoriously have ears, and they are very cagey on this island.'

'What about the girl? You were so lovey dovey.'

'I'm still lovey dovey, as you put it. She's not important; just a pawn in our game.'

'I don't know whether we can trust you, Caramba.'

'I am Romeo Curria's son, Muso; never forget that.' James' words sounded curiously threatening.

Muso Medusa visibly curbed a barbed reply and nodded. 'OK., Caramba, no hard feelings. Your Dad is on the island and this evening everything will be decided.'

'That's a surprise,' James said. 'Where is he?'

'Can't tell you that, but come here around 6pm and we'll go to see him together.'

James drove back to the club and asked to see Maurizio. He was told that Mr.Capponcini would not be available until the late afternoon. James went to his cottage and put a call through to Jason Reed. He told Jason about his meeting with Muso Medusa and that he was going to see Romeo Curria, who was supposedly on the island, in the late afternoon. Jason agreed that James should go and report afterwards to Maurizio at the Club.

James hung up and thought about how he was going to get through the day without seeing Miranda. He might just catch a glimpse of her in the grounds of the Club, but had a feeling that she would keep out of the way. He went out onto his porch and looked at the bay. The catamaran wasn't anchored there any longer. He wondered whether the Capponcinis had gone out on a sail. He felt lonely and listless. He decided to go and look at the bulls again and try a few veronicas.

The catamaran cruised along in the brisk breeze, and Maurizio held the wheel steady. The salty spray of the waves mingled with Miranda's tears as she expertly handled the sails. Domitilla looked at her daughter anxiously. She had never known her to be so tearful.

'Where shall we stop for lunch, ladies?' Maurizio called to them.

'How about Bugler's bay? They might have brought in a lobster,' Domitilla answered. 'Is that OK with you, Randy?'

Miranda nodded silently and Maurizio steered towards the calm waters of the bay in front of them. Miranda pulled in the sails and Maurizio turned on the engine. He carefully negotiated the passe and Miranda threw out the anchor. The tiny beach was almost deserted, and the restaurant had the red flag up. The flag proclaimed that the innkeeper had successfully raided the lobster pots in the bay. The Capponcinis climbed into the dinghy and made for shore. It wasn't long before they were demolishing the steamed lobsters. Only Miranda toyed with the carcass on her plate, staring dejectedly out to sea.

'Randy, you've got to snap out of this. We've told you what the plan is and James is coming for dinner tonight to keep up appearances.'

'I'm going out then.'

'Don't be so stubborn, Darling,' Maurizio admonished his daughter.

Miranda frowned at her parents. 'Alright, so I won't go out. I'll stay in my room.'

'If that's what you want to do, we can't stop you. Now do eat something, Darling. That's a really good lobster; it's always been your favourite.'

'I wish I could make you understand how confused I am... I feel so terribly let down and at the same time I love him so much. I want to be with him so badly, but I can't forget the real reason for his presence in Tequila. I'll never get over this.' She looked at her father. 'And don't try to hide that smile of yours, Papa. This is really serious.'

'I know, Darling, truly; I know that it's serious. It's serious for James as well, I'm certain of that. He is in a most unenviable position, and is trying to extricate himself in an honourable manner.'

Miranda tossed her mane. 'You don't need to be too concerned for him, Papa. You always told me that blood is thicker than water. A Curria, that's who he is.' Miranda stabbed into the lobster vehemently with her pick, and dislodged some meat. She looked at it and put it down on her plate. 'I'm sorry; I just can't eat this. I'll go for a swim until the two of you have finished,' she said and left the table.

James looked over the animals carefully, as he sat on the fence which ringed the bull pen. He was always meticulous in his preparations for the fight. This time however, he found it hard to concentrate on the matter in hand. He was restless and anxious. It was a condition he wasn't accustomed to. The thought that both his mother and his father were to be within a few hundred square feet of each other on Tequila was brutal. His mother would probably arrive tomorrow on the same flight he had come on and be closely guarded at the US consulate. Only Muso Medusa knew where Romeo Curria was hanging out. Curria was a threatening figure who was going to demand that James fulfil his obligations. James suddenly felt very young and vulnerable, and abandoned on this island. He jumped off the fence and went into the ranch house where lunch was being served up to the ranch-hands. He was glad of their noisy invitation to join them, and he ate hungrily. He hadn't had dinner the night before, or breakfast in the morning, and the steak and chips smelled good and tasted better. Nevertheless, he couldn't join in their cheerful good humour, and left as soon as he had cleaned his plate. He knew where he would find comfort and advice, and perhaps even a look into his future. He would pay Sorciero and Old Crone Chang a visit. He stopped at the petrol station and asked direction to Sorciero's house in the forest. Everyone knew where Sorciero and Old Crone dispensed their magic, and before long James had reached the dirt track which led to the cottage.

The girl sitting at the table on the porch yawned, and looked at the new arrival. It was almost the end of the afternoon and she was ready to go home. It took her only a split second to recognize Caramba De la Cruz. Instinctively she smoothed her hair and wet her lips. It wasn't every day that such a cool customer came to the cottage. He was just as gorgeous as his photographs and she felt weak at the knees when he approached her with a friendly smile and asked whether Sorciero could receive him. The girl grinned wordlessly and went into the cottage, waggling her hips as she went along. She returned, and held the bead curtains open for him. A sweet smell of incense hit James' nostrils as he entered and squinted into the gloom of the consulting room. Old Crone Chang was perched on the Mandarin throne, and Sorciero, in full witch-doctor's regalia, was sitting in front of a crystal ball. They formed a strangely impressive duo, the old Chinese woman and the Shaman of Tequila. Coal embers were glowing in a bronze brazier, and little baskets of various herbs and apothecary bottles containing coloured liquids, were arranged on the shelves.

'Welcome, James De la Cruz' Sorciero greeted him. 'I had a feeling that you might be coming to see us.'

'What can we do for you, young man?' Old Crone said. 'Would you like a cup of my special tea, before we begin?'

'That would be cool, Ma'am. I've heard good things about it,' James replied politely. Old Crone got up from the Mandarin throne and put the kettle on to boil. She added some leaves into her pot and poured the water over them. She sniffed the delicate perfume and when she thought that the leaves had drawn enough she poured the liquid into one of her fine china cups. She offered the cup to James.

'Sit down, young man, and drink the tea,' she commanded.

James obediently sat down in front of Sorciero and took a sip of the tea. He couldn't control a grimace at the taste of it. 'It improves as you drink it, James; trust me,' Sorciero said and smiled. 'Now tell us all about it.'

James stared morosely into the tea cup, then took another sip. This time the tea tasted better. 'It's about my Mother. She's supposed to arrive in Tequila very soon, so that we can look after her.' James looked up. 'You know all that, but there is more. I'm worried because it seems that my father, Romeo Curria is on the island. I'm supposed to meet with him later on this evening.'

'No kidding. That's tough shit, Mun.'

'I don't know quite what to do. Can you look into that crystal ball of yours and give me some advice?'

Sorciero nodded and gazed intently at the crystal ball on the table. He waved his hands over it and muttered some incantations.

'This shit don't work for me, Old Crone. What does the Mère Magot know that I don't? I spent a small fortune on this Baccara ball today, and it don't tell me a thing.'

'Patience, Sorciero; you need a little time to practice. It will come to you as did all the other magic, because you are a true witch-doctor,' Old Crone pacified him.

Sorciero continued to wave his hands about and suddenly gave a little cry and pointed to the ball which seemed to have taken on a radiance all it's own.

'Look, Old Crone, it's working,' he jubilated. 'The blasted thing really works. It's throwing fireworks all over the place.'

'What do you see, Sorciero?' James asked with a trembling voice.

Sorciero stared and stared, concentrating all his mental capabilities on the crystal ball.

'It's a strange thing, Mun, but I don't see your father on the Island, but I can see your mother; she's a pretty lady, with dark curls and a good body. Yeah, long legs. You don't really look like her, with your fair hair and light eyes.'

'I don't look like my father either. He's dark and black eyed. I saw a photograph once.'

'There's a dark, black-eyed man here, but he's not your father. Maybe it's the Medusa fellow. You're in deep shit there, Mun. He's expecting you to do their dirty work.' Sorciero murmured and continued to gaze at the crystal ball. 'There's going to be big trouble... Oh my God, it's Charlie and he's crying.' He started up, and looked around wildly. 'I gotta go home, Old Crone, right now.'

'I'll come with you, Sorciero,' Old Crone said. 'Charlie'll be alright, there must be some mistake. You didn't interpret the image correctly.'

'I'll come too.' James cried.

'No, young man, you must go and keep your appointment with Muso Medusa,' Old Crone said. 'Anything else will arouse their suspicions.'

The two cars drove down the dirt road and shot onto the tarmac road. James gave a short wave to Old Crone and Sorciero as they turned off at the main highway. James continued down the road into the town. There seemed to be little point in going back to the club for a half hour. He stopped at a side walk café, and ordered an espresso. As usual, his appearance almost stopped the traffic. The café seemed to fill miraculously with a bevy of young women, who covertly ogled the handsome Torero. He gulped the espresso, burning his tongue in the process. Cursing silently he went back to the jeep and drove to the Carlton Hotel. He was filled with apprehension, almost as if he were about to step into the bullring. In the arena, at least, he had a fighting chance. He could stare the bull in the eye and instinctively know what the animal's next move would most likely be. In the Lion's den he was about to enter, his weapons of self defence were nonexistent. He pulled up outside the entrance and called the bellboy. He asked him to tell Mr. Medusa that Caramba De la Cruz was waiting for him.

Muso Medusa came out, dressed in candy striped pink trousers. His stomach bulged alarmingly over the tight belt. A loose-fitting baby blue linen jacket completed his outfit. James shuddered at the unsightly corpulent mass which marched towards him.

'Get out of the car, Caramba; we're going in my limo,' he said.

James was about to demur but thought better of it. He gave the keys to the bellboy and asked him to park the vehicle. Muso waved at a limousine waiting under the canopy. The chauffeur brought the car up and sprang out to open the door. Muso and James got into the back of the car and were driven to the exit. A few miles up the road, Muso leaned forward and tapped the chauffeur on the shoulder.

'Stop the car, Puzzo,' he said.

'Si, Signore,'

'Get out, Caramba, please.'

James shrugged and obeyed.

'Frisk him, Puzzo,' he ordered the driver.

'I'm not armed,' James protested and pushed the driver away.

'If you're not armed, then why do you object to being frisked? Nobody goes into the presence of Romeo Curria without being frisked.'

'But I'm his son, damn it. Would I kill my own father?'

'It's been known to happen in the best families, so stop wasting time. Romeo Curria doesn't like to be kept waiting.'

'Oh, alright,' James said and submitted to the driver's investigations. When Muso was satisfied that James was clean, he produced a bandanna from his deep pocket.

'I'm going to blindfold you, Caramba,'

'No, never; I positively refuse. I get panicky in the dark.'

'Rubbish; don't resist, there's a good boy.'

They drove for a long time it seemed to James, sitting in the dark. He tried to orientate the way the car travelled. It seemed to be going round in circles. It stopped at last, and James got out and was made to climb a rope ladder down a steep embankment. He heard the splash of water against a jetty, and then was walking up a plank onto what was surely the deck of a boat. Nobody tried to stop him when he ripped off the bandanna. It was already dusk and he could just see part of the small bay where the yacht was moored. James took a quick peek over his shoulder and saw that the bay and the hills behind it were dark and deserted, except for the sea swallows which were agitatedly trying to protect their nesting places in the rock-face from the intruders. Obviously it was a part of Tequila which had escaped development. James turned his attention to the yacht he was on. It seemed a compact, luxurious vessel, smartly kitted out with appropriate deck furniture in yellows and blues. The deck lanterns were on but dimmed. A shiny mahogany bar, trimmed with brass, graced the stern. A steward stood behind it. Two deckhands in starched whites stood at the side. Their ample Polo shirts couldn't quite hide the slight bulge of the guns they had tucked under their arms.

'Tell the Capo that we've arrived,' Muso commanded the deckhands.

'He said for you to sit down and wait up here, Sir.'

'I've got to use the head,' James said. 'You've been driving me around for hours, and I've got to go.'

'You'll just have to wait, Caramba,' Muso said unsympathetically. James made to undo his trousers, and the two deck hands almost pulled out their guns.

'OK., I'll just piss all over this immaculate deck. Maybe I can aim for that lovely yellow linen armchair,' James said cheerfully.

'Take him down to the guest head, and don't leave him alone for a second,' Muso growled at the deckhands, who had put their guns back. One of them led the way down the companionway to the staterooms. All the doors were closed and the sound of a television came from one of them. It sounded like a children's programme. The deckhand opened the door to the head and motioned James in. He turned to close the door, but the man shook his head. James unceremoniously pushed the man back, banged the door and locked it. He could hear the television more clearly now. He could recognize the 'Postman Pat' series which he had watched on video when he was a child. There must be some children on board the yacht, and he wondered who they could be. Maybe some half siblings of his; that thought didn't really please him. He opened the porthole and stuck his head out. He saw a small dark face glued to the closed porthole next to his. The child was mouthing something unintelligible and banging his little fist on the porthole. James waved back, and looked further. There was nothing to see except the stars now beginning to show in the clear sky. It was very quiet;

only the sound of the sea lapping against the side of the yacht disturbed the night. He closed the porthole and wondered who the child was. He used the head and flushed it. He unlocked the door and came out. An irate deckhand was standing in front of him, arms akimbo. The sound of Postman Pat had been increased in volume and he thought he heard the child crying.

A scantily dressed young woman came out of one of the staterooms and stopped to stare at James. A grey-haired elderly man in boxer shorts came out after her and wanted to draw her back. She tried to resist, and tossed her long blonde mane. Suddenly the man slapped her hard across her face, and she whimpered in pain. James was tempted to come to her aid, but the feeling of something cold and hard in the small of his back which felt uncommonly like a weapon, made him change his mind. It was after all, none of his business. The deck hand beckoned him up the companionway and James obediently climbed up on deck. Muso Medusa had helped himself to a rum punch, and offered James a drink. James accepted a glass of mineral water from the steward.

'Where is my father?' James asked impatiently.

'He knows we're here. He's just getting up from his nap and will join us when he's ready,' Muso said. 'Besides, what's the hurry? Relax and enjoy the feeling of luxury this beautiful yacht radiates.'

'I can't stay here all night,' James complained. 'I have a date for dinner with the Capponcini girl.'

'She'll wait. Train your women right from the beginning. Waiting is their lot. Waiting in more ways than one. Waiting for their men and waiting on their men; that's what our women must do. Your mother waited, didn't she?'

James felt his blood starting to boil. He would have liked to plant his fist into Muso Medusa's smug, ugly mouth. He remembered Jason's instructions and contained his temper. He settled down on one of the comfortable deck-chairs and made the ice in his glass clink.

'If you say so, Muso. It's no skin off my nose; but don't forget, she's not been used to Mafia manners, so she might not wait.'

'She's Italian, isn't she? They always wait. It's inborn, my boy; you take it from me. I'm an old hand at the game.'

James wanted to blurt out that he didn't want to take a damned thing from Muso, except the knowing smile off his face. He turned away and continued twiddling with his glass. The deckhand, who waited at the companion way, suddenly stood to attention. Firm steps could be heard coming from below. James recognised the man he had seen at the door of the stateroom. The tousled grey hair had been smoothed down, and the soft white linen jacket concealed the pot-belly. Romeo Curria projected an aura of authority and elegance as he stood on the deck of his yacht, and let his gaze sweep across his visitors. The sight of his son James pleased him. He was happy to have conceived such a beautiful young man, who also had the reputation of being courageous. A true torero, who entered the bullring fearlessly. Romeo Curria strode towards James, who put his glass down and got up. Curria embraced him and kissed him on both cheeks.

'Figlio mio, I am glad to meet you at last,' he said.

'What took you so long, I've been waiting a long time,' James said reproachfully.

'I had to take care of business, my son.'

'For twenty-one years? That must be a hell of a lot of business,' James said and loosened himself out of the embrace.

'You are right, my son, it has been too long and I am sorry for that. Now we will make up for lost time.' He lifted his finger and the steward brought over a glass of rum punch. 'First we have to discuss business, then we will take our pleasure.'

'Business is fine, but I'll take my pleasure elsewhere, Sir,' James said.

'Call me Papa, figlio mio, I would like that.'

'I would like that too, but it's too soon, Sir; you can understand that. In time perhaps, I'll be able to call you Papa.'

Muso Medusa made to speak, but Romeo Curria shut him up. 'He's right, Muso, we must get to know each other first. He does not know me, but I always cherished him. His dear mother always sent me photographs and wrote me letters about his progress.' He smiled benignly at James. 'So tell me, my son, how far have you got in your mission?'

'Well, I decided to stay at the Club Royale and made contact with the Capponcinis. I have taken their daughter Miranda out, and she's crazy about me. In fact, we're like that,' James smirked and linked his little fingers together. 'I'm invited to her place for dinner again tonight.'

'Excellent work, figlio mio. Now listen good; before the week is over, we have plans to take over Petrolia. I have decided that my interest in Tequila is limited. The elimination of the President, her buffoon of a husband, the old Chinese trout, her daughter-in-law, and the Capponcinis is irrelevant. They can keep their Tequila. Without Petrolia, Tequila will return to being a Banana Republic, just as it was before. We hope to take over Petrolia by peaceful means, but are prepared to invade it, if necessary.'

'That's cool, Sir,' James managed to put a reasonable amount of enthusiasm into his voice. 'But by peaceful means? I don't understand... How are you going to achieve that, Sir?'

'Leave that to me, my boy. Your duty is to get yourself taken to Petrolia for a visit and find out how they are protecting their oil wells. Get the girl to arrange a little picnic there, or something.'

'Hmm, that sounds like a good move, Sir. Time is running out of course. It's almost the beginning of the Semana Santa and everyone is very busy. A bit short notice, I'm afraid.'

'Do it tomorrow.' The words sounded uncommonly like a command, and James felt the hair on his neck bristle in antagonism.

'Yes, Sir,' he said then. 'I'll do my best.'

'Report to Muso, and he'll keep me informed. You mustn't come back here again. No one must know of this location.'

'Yes, Sir,' James decided to add. 'You won't be disappointed in me, I promise.'

Romeo Curria nodded with satisfaction. He dismissed Muso Medusa and James with the wave of a hand and went into the air-conditioned saloon. The bandanna was tied over James' eyes again and they climbed up the rope ladder to the top of the cliff. They sat in the car in silence, until after an hour of twists and turns, the limousine drew up in front of the hotel. The bandanna was removed, and James heaved a sigh of relief. He waited for his own vehicle to be brought around and put his foot on the gas, as soon as he had left the hotel grounds.

There was only a little time left for him to change, and he phoned 'Tradewinds' and apologized for the delay. Fifteen minutes later he had showered, changed and picked up the roses he had ordered in the club flower-shop in the morning. He drove the jeep to the house in record time. The vehicle seemed to know the way by itself as it snaked up the gravel road and stopped at the door. James got out, gathered the flowers and rang the bell. His heart started to beat a wild tattoo at the thought of seeing Miranda again. He hoped that she would forgive him, but only time would tell.

Chapter Eight

Sorciero and Old Crone didn't speak as he drove the car at breakneck speed towards the presidential villa. He couldn't get the vision of Charlie crying and asking for help out of his mind. Old Crone tried to activate his mobile phone, but he had forgotten to charge it that day, so there could be no incoming calls either. He cursed his absentmindedness and felt helpless and close to tears. Maria Beatriz was in the middle of a social engagement. It was the big Easter Egg hunt fete for the Arts & Crafts Society Benevolent fund; she was the President of that charitable organisation, and it traditionally took place in the gardens of the presidential villa. A great number of mothers with their children came to the fete. The entrance fee was five dollars for the family, which went into the kitty for the funds. The President provided an enormous tea with all the trimmings, and the children were busy looking for the chocolate eggs and other little presents which had been donated by the leading businesses on the island and hidden in the grounds of the property.

Cars were parked all down the road leading up to the gates, which opened to let Sorciero's jeep in. The gardens were milling with excited children and their mothers, all intent on finding the goodies. A big buffet table had been set up on the lawn, and the chef was barbequeing a mass of sausages, sweet corn, and chicken kebabs. Sandwiches and little fancy cakes, potato crisps and cheese straws were laid out on the billowing white cloth, ready to be eaten. Maria Beatriz, looking splendid in an embroidered emerald green caftan, with matching turban, was surrounded by guests. Sorciero drove round to the garages and he and Old Crone hurried across the lawn to Maria Beatriz.

'Hello, Sam, I'm glad that you could make it early. Lovely that you could come, Old Crone.' Maria Beatriz beamed. 'The children are so excited, they're having a wonderful time.'

'Where is Charlie?'

'Oh, he's around playing with the children. His basket was already full with goodies.' She looked at him and saw the worried frown on his face. 'What's the matter, Sam? You look sick.'

'So would you look sick, if you had seen what I saw,' he muttered. 'When did you see last see Charlie?'

Maria Beatriz wrinkled her nose and thought. 'It's just a little while ago. He was with Ling-Ling and some other children; they were playing hide and seek, or perhaps they were on the swings at the back of the villa. Why don't you go and have a look there? It's time for tea anyway, and it'll be getting dark soon. I'll tell the gardener to put the all the fairy lights on and you bring all the kids around for tea. There are going to be fireworks afterwards.'

'I think I'll stay here with you, Ma'am,' Old Crone said. 'Perhaps I can have a word in private with you?' Maria Beatriz nodded and the two women left the group of guests.

'What's up with Sam? Did you have a bad day at your sessions?'

'Well not really, Ma'am. It's the new crystal ball he bought down at 'Little Switzerland'.

'It was very expensive, I know, but then he wants to be able to learn new things. The Mère Magot certainly was quite impressive when she used hers. Frankly though, I'm not convinced,' Maria Beatriz said.

'I think Sorciero saw something in his new crystal ball. Something which really frightened him,' Old Crone said softly. 'He has a very vivid imagination, that's part of his job, but I think he genuinely saw something today. I don't want to alarm you, but it was something to do with your son.'

Maria Beatriz clutched at her bosom. 'My God, what's happened? Where is Charlie? I was so busy with the guests, I didn't notice the time go by. We must find him.' She started to walk quickly across the lawn around to the back of the villa. Children were queuing up for a ride on the swings, watched by their mothers and nannies. Sam was running around looking everywhere, asking the mothers and children whether they had seen Charlie. Most shook their heads, but a little girl piped up and said that he had been taken for a pee ages ago. She put her hand up to her mouth and giggled.

'Thank God; he must be in the house, then,' Sorciero said. They walked quickly to the back entrance and went through the kitchens in to the main hall. There were guest cloakrooms there and they called out for their son, but there was no reply. They threw open all the doors, but found no one there. They ran upstairs, and into the nursery. They stopped aghast at the sight of Ling-Ling lying on the floor, hands and feet bound, and gagged. A big pad of cotton wool was lying beside her. There was a distinct odour of chloroform in the air. She was struggling to rid herself of her bonds, but the aggressors had done a good job. Sorciero liberated her and she coughed and sputtered, trying to catch her breath.

'What happened Ling-Ling? Where is Charlie?'

The girl started to weep and could hardly speak. Old Crone Chang spoke to her sharply in Cantonese and the girl managed to control her sobs.

'We were attacked, Ma'am, and they pressed these cotton wool pads on our noses. When I woke up, I was all tied up and Charlie was gone.'

'Who were they, girl? Did you recognise them?' the girl shook her head.

'They were wearing black hoods, and it went so fast, that I'm not even sure how many there were.'

'There were more than one?' Old Crone asked.

'At least two, Old Crone.'

'Men?'

'Yes, it seemed to me they were men, although I don't know... one of them spoke and sounded more like a woman.'

'They spoke English?'

'Yes, Ma'am, and I heard an Italian word.'

'Island English?' Sorciero wanted to know.

'No Sir, it sounded American.' The girl sat up and rubbed her wrists.

'We must go and find him, Sam; he must be so frightened, our poor little boy,' Maria Beatriz cried and hid her face on Sorciero's shoulder.

'We must telephone Chief Pereira at once,' Sorciero said. You must go back to the party, Sweetness. Nobody must know what's happened. Old Crone and I will start the search.'

Sorciero found the telephone in the hall and was about to dial, when it rang. They all stopped in their tracks as if turned to stone. It rang several times before Sorciero finally picked it up.

'President's Residence,' he said, and waited. 'Yes it's me, Mun, Sam.' They could hear a torrent of words burst through the receiver. Sorciero shook his head in despair and said: 'Yea, Man, I hear you. I'll follow the instructions, don't worry,' and put the receiver down. He turned to the three women.

'It's them, the kidnappers. They've got Charlie and he's safe they say.'

'How much, Sam? How much do they want? They can have everything we've got,' Maria Beatriz moaned.

'It's not money, Sweetness; Oh God, it's not money.' His face crumpled and he wrung his hands. 'They want Petrolia in exchange for Charlie. They have given us thirty-six hours to agree and sign the document which will transfer ownership of Petrolia to them.'

'We must call a meeting with Jason Reed and the Capponcinis. You call them Sorciero, and I will get Mei-Ling. It's best if we all go to "Tradewinds".' Old Crone had taken the situation in hand, and with her ubiquitous basket swinging from her fragile arm, she led the way out of the presidential villa to the parked cars.

James Harvey de la Cruz stood on the 'Tradewinds' deck, holding a glass of iced tea. Tonight he would have liked to have a real drink, to feel the alcohol burn right down to his entrails. He remained disciplined however; no alcohol before a corrida. He heard the telephone ring, and Maurizio went to answer it in the sitting room. James could hear Maurizio's voice raised in agitation, and saw Domitilla catch hold of his hand. There was no sign of Miranda. Perhaps it was she, announcing that she wasn't coming in for dinner. Maurizio and Domitilla came out hand in hand. Their faces were pale and grave.

'What's wrong, Sir? Has Miranda had an accident?'

'No, James. She's safe, thank God. She won't budge

from her room though. It's Charlie, the President's and Sorciero's son. He has been kidnapped, and is being held for ransom,' Maurizio said quietly. 'Sam, Jason Reed, Mei-Ling and Old Crone are on their way here. We have to take some kind of action. Sam couldn't tell us everything on the telephone, but it sounds pretty bad,' Maurizio continued. He looked at James. 'You said that you had some news for us?'

"Perhaps it had better wait until they're all here, Sir.'

'Should we dine now, Maurizio? I have to tell cook if we postpone.'

'Might as well, although I'm not really hungry. Are you, James?' James shook his head. 'I'll have some more iced tea, please.'

'You may not feel hungry but we must all eat.' Domitilla turned to go to the kitchen. 'Do sit down; everything's sure to be ready.'

They made short shrift of the home-grown tomato salad and excellent sea-food risotto. James kept on hoping that Miranda would relent and join them at the dinner table, but he was disappointed. Cook came to clear the table and put out the sliced pineapple for dessert on the sideboard. They heard the jeep scatter the gravel as it came to an abrupt halt outside the house. The visitors came out to the deck, Sam leading the way. Domitilla poured out the iced tea, and ordered some hot water for Old Crone from Cook. Maurizio asked Sam to tell them in detail what he knew. When they all had listened to him, he opened the shopping bag he had brought in and took out the crystal ball. It glowed in the subdued lights of the deck.

'Fine ball, Sam. Looks like a Baccarat.' Maurizio said. 'You say you saw Charlie in the ball this afternoon, and that he was crying.'

'That's exactly what happened, Mun. James was there at the time, he can vouch for that. I knew something was wrong then, and Old Crone and I sped back to the villa, and found that Charlie was missing.'

'Well, try looking into the ball again, Sorciero. There may be more for you to see.'

Sorciero nodded and sat down in front of the ball. The others stood behind him, hoping to see something as well. Sorciero started to murmur tonelessly and waved his hands over the ball. At first there was no reaction, but after a few minutes there were faint flashes of light coming from the ball. Sorciero looked over his shoulder and asked the watchers to move away. They complied and gathered at a little distance in front of him. The ball started to glow more brightly, and suddenly Sorciero gasped and gripped the edge of the table. 'I can see him; I can see Charlie. It looks as if he's on a boat.' He looked up at the others and cried: 'We must go to the harbour; they're keeping him on a yacht, for sure.'

'Can I interrupt?' James asked. They all turned to him expectantly. 'I'm almost sure I saw Charlie this evening before coming here.'

'Where, Mun, for God's sake where?' Sorciero's voice trembled with excitement.

'I can't tell you exactly where, but I'll try. Muso Medusa took me to see my father, Romeo Curria, tonight.' He looked at them grimly. 'They drove me blindfolded in a limousine round and round Tequila. Then I climbed down a rope ladder attached to a steep cliff-side and ended up on the deck of a luxury yacht. Romeo Curria's yacht. I tried to look around, but it was almost dark, and only some sea birds nesting in the cliffs were swooping around. No lights anywhere.'

'Get on with it, Mun; where did you see Charlie?'

'I saw a child waving to me from a porthole. About 4-

or 5-years old. He was dark, with a lot of black curly hair. It might well be Charlie. Your crystal ball says a boat, and Romeo Curria has already admitted to having managed that abduction. He wants revenge, as I told you. He ordered me to get a permit to visit Petrolia tomorrow, and check out how the island is guarded.'

'So he wants to have more than one option,' Jason Reed commented.

'Describe the place where they took you, again,' Mei-Ling asked James.

'As I've already told you, I was blindfolded for most of the time. It seemed to me that it was a deserted stretch of water, probably a bay, as the water was calm. A bay surrounded by steep cliffs and nesting birds. No lights anywhere on land.'

'It sounds like Little Bay,' Mei-Ling said. 'Of course it could be somewhere else, Lobster-Pot perhaps; but Lobster-Pot has steps carved into the cliff face. You sure that you heard birds, James?'

James nodded. 'Quite sure.'

'Then I'm fairly confident that it's Little Bay.'

'No one else knows the coast line as well as you, Mei-Ling, but I thought that the passe was too narrow for a big vessel.'

'Not if you know it well, Maurizio. We've always put the word about that it is too narrow. It says so in the guide books too. It was done for conservation, so that the sea swallows which nest there, wouldn't be disturbed. I do have a buoy there for my dive boat.' Mei-Ling pursed her lips. 'It's not easy to find out if there's a yacht anchored there. Even though the passe is so narrow, Little Bay is quite wide, and a yacht anchored in the far corner wouldn't be seen by other passing yachts. Chief Pereira could send out the coast guard, or send the helicopter out for a reconnaissance.'

'We could storm the yacht in the night, and get

Charlie out,' Sorciero said.

'But if Charlie really is on that yacht, and their suspicions were aroused, they might harm the child, before we have a chance to board the vessel. They are bound to have sentries on duty twenty-four hours a day,' Jason commented. He looked at James and raised his eyebrows questioningly. 'Tell us everything you saw on that yacht, James; how many people were on it. Every single thing that you can remember.'

James nodded and thought for a moment. Then he recapitulated all that he could recall. The steward behind the bar. The two deckhands, both armed. His visit to the lower deck, where the cabins were. The child he had seen at the porthole. Romeo Curria in his underwear, slapping a young woman around. He had seen no one else.

'What did this young woman look like?' Old Crone asked. James could hardly suppress a grin as he answered:

'She had long blonde hair. It reached down to her bottom, which seemed ample to me. Perhaps not to everyone's taste, but quite sexy,' James mused. 'She never said a word though, so I don't know if she was Italian or what.'

'I don't know what you think,' Old Crone turned Jason, 'but it sounds to me that this girl is none other than Maria Dolores. She could have taken up contact with the Curria clan in New York and they could have plotted their vendetta together. They used her first, hoping to get young Charlie to go with her, and when that didn't work, they applied the strong-arm method.

'The girl obviously likes older men,' Mei-Ling said sarcastically. 'Now that they've got young Charlie, I don't suppose that she will leave the yacht again. The danger of someone recognizing her makes it very dangerous for her to be seen.'

'Did they have someone standing guard at the rope ladder at the top of the cliff?'

'I don't know. They didn't take off the bandanna until I was on deck. By the way, I did see the name of the yacht painted on one of the life belts.'

'Good man; what was it?' Jason praised him.

'Sharkstooth. BVI.'

'Very apt. So it's registered in the British Virgin Islands. I'll get Chief Pereira to find out who the owner is. May I use your phone, Maurizio?'

'Be my guest. It's in the hall.'

Sorciero looked around at the others. 'We could climb down the rope ladder in the dark and take the boat by surprise,' he said eagerly. 'We could try right now; Little Bay is only a few kilometres away. What do you say?'

'You can't take the car all the way to the edge. You must go there on foot, as quiet as a mouse.' Mei-Ling looked at them. 'Who's coming with us?'

'Don't go, my daughter. Remember that you are with child,' Old Crone protested.

'I'll be careful, I promise. I've got to help them. I know these little bays better than anyone.'

'I'll come with you,' Maurizio said. 'I think James and Old Crone had better stay here, with Domitilla. Sam, you must telephone Maria Beatriz and tell her that we might have found Charlie, and that he appears to be unharmed.'

'What about the police? Should we tell Chief Pereira that we might be taking things into our own hands?'

Jason came back to the deck. 'I've spoken to the Chief, and asked him to check the registration,' Jason said. 'In the meantime, let's find out if the yacht is really anchored in Little Bay.'

'Are you armed, Jason?' Maurizio asked. The Consul nodded and patted his back pocket.

'Any one else got a gun?'

'I've got one, as you know.' James said. 'I left it in the jeep.'

'Good, then we can take that one too.'

'Mr. Reed, before you go, can you tell me anything about my mother?' James asked.

'She is already in a safe house in Buenos Aires. We hope to have transferred her here by tomorrow evening. She is doing fine, so don't worry.' James managed a smile of thanks, but his forehead was still creased in a deep frown.

'Why can't I come with you?' he asked. 'I feel totally useless hanging around here.'

'No way. If anyone spotted you there, it would be disastrous for you.'

Mei-Ling drove the land-rover down the secondary coastal road. It was a road seldom used now. The new tarmac two-lane highway had absorbed most of the island traffic. She drove slowly, carefully avoiding the potholes. She dimmed the headlights, and then shut them off completely. She put her finger to her lips, and motioned them to leave the vehicle. She took a small revolver and a torch out of her purse and joined the three men. The stars were out, but the moon hadn't risen yet and it was dark. They followed her down the road, treading lightly, trying not to disturb the gravelled surface. She held up her hand and pointed to a path which led through the bushes towards the sea. They could hear the surf crashing against the outer cliffs. She threaded her way through the greenery, as quiet as a mouse, waiting after each step to listen. Suddenly music wafted up; it was an aria from a Puccini Opera. She clapped soundlessly and knew that they had hit the jackpot. There didn't seem to be anyone on guard on the edge. Mei-Ling looked over the edge and saw the boat. It had tied up to the old abandoned jetty. A local fisherman had built himself a shelter there years ago. Now all that remained were four posts, and the remnants of a thatched roof.

Mei-Ling retreated soundlessly towards the bushes and joined the others They came out onto the dirt track, and waited for her to say something.

'The yacht's there alright, and I could see some people on the deck. It must be the same boat our James was on. It would be too much of a coincidence. Besides, they were listening to Opera. Mafiosi always listen to Italian Opera.'

'What do we do next?' Sam asked anxiously.

'They've removed the rope ladder, so we can't get down there now. It might be too noisy anyway. Bits of cliff might crumble and fall and give us away.' She looked at Maurizio and smiled.

'Oh no, my dear girl, I have no intention of letting history repeat itself,' Maurizio said sharply. 'No night diving for me, or for you. You're pregnant, and it's much too dangerous.'

'It's still going to be the only way, my friend. Or can you think of another plan?'

'I think that we must really gather all the facts first, and then plan a move with police back-up,' Jason said. 'We must be absolutely sure that it's the right yacht. Opera music isn't really enough to go on.'

'Come on, Mun,' Sorciero objected. 'It's a sure thing. Charlie is on that yacht, and I'm going to get him out.'

'Be reasonable, Sam. We need help.'

'So let's get it right away. Let's go back to "Tradewinds" and organize it,' Sam urged. 'My instinct tells me that now's the time to strike those bastards. It will come as a complete surprise if we act soon.'

They sat on the deck at 'Tradewinds' sipping the last

of the iced tea. Sorciero was busy with his crystal ball, and kept on saying 'Yea, Mun, yea.' Jason was on the telephone to Chief Pereira who had been hauled out of his bed. A coast guard cutter was organized with a couple of divers Mei-Ling had trained. James wanted to go on the dive with them, to show them which porthole he had seen the child at. Sorciero would be in the dinghy with the patrol boat trailing behind. The patrol boat was to come as close to Little Bay as possible and then drop the divers off. They would reach the yacht undetected and hang about there. After a short time the coast guard vessel would enter the Bay, and accost the yacht peacefully. A routine check as it were. The coastguards would keep the crew and the owners busy with questions, look at the ship's log, and generally distract them and keep them up top. In the meantime, the divers would find the child and bring it safely to the dinghy which was tied to the coastguard vessel. At the same time, police were to be stationed at the top of the cliff, to prevent any one trying to escape from the yacht. If all went according to plan, nobody would even have noticed that the boy had gone, until they were well away.

'It's a good plan. Let's hope that there will be no bloodshed,' Old Crone murmured. She looked at Jason. 'But what about the culprits? How are we going to deal with them?'

'That will be my business, Old Crone. I will put a headache on them they'll never shed for the rest of their lives. No need for guns or grenades. Just my little wax figurines and a few pins, and they will wish they had never come to Tequila.' Sorciero vowed.

'Now we just have to wait for the patrol boat and hope that the Chief can get his act together.'

Around midnight, the patrol boat entered the bay at

'Tradewinds'. Sorciero and James were down on the jetty with their equipment, which was quickly loaded. They also took blankets and some pullovers to wrap up in after the dive. Mei-Ling spoke to the two coast-guard divers, and briefed them patiently. She sensed that they were nervous, and decided that she would join them aboard. She ran up the steps and embraced her mother-in-law and Domitilla. She promised faithfully that she wouldn't dive, but explained that the divers were not very experienced, and would need her expertise and moral support. Maurizio was to take the land party to the cliffs above the bay and stay there with them, until the operation was completed. Old Crone and Domitilla would stay at 'Tradewinds' and listen for their phone calls. They all had their mobiles with them, and would report back as often as they could. In turn, Old Crone and Domitilla would keep Maria Beatriz informed.

The patrol boat manoeuvred out of 'Tradewinds' Bay and made for Little Bay. The sea was relatively calm, as they rounded the steep cliffs of the Atlantic side of the island. The captain cut the engines and they drifted with the current towards the passe. The anchor was let down, and the divers prepared for the dive. Mei-Ling checked all the bottles, and made them check their BC's, and the regulators. One by one, they let themselves glide into the water and submerge. The lead diver had a line attached to his belt, which he would pull three times to let the patrol boat know that they had entered the bay. He would then pull on the line twice to let them know that they had reached their target. Repeated tugs would signify that there was big trouble ahead, and the divers were returning to the patrol boat.

By now it was 1 a.m., and a few clouds had dimmed the stars. It might even come down in a shower, as it so often did in the Caribbean. Mei-Ling stood on the deck with the ships infra-red binoculars pressed to her face. She could just see the bubbles of the divers as they approached the passe. She felt the three tugs and knew that they were in the bay. She nodded at the captain, who stood beside her. They were about to start operation 'Get Charlie.'

Suddenly there were repeated tugs on the line. The captain and crew were galvanized into action. Mei-Ling stared through the binoculars and saw the bubbles advancing towards them. Within minutes, the divers were climbing up the ladder. The lead diver tore off his mask and gasped: 'The yacht's gone; as far as we could see, the bay is empty.'

Sorciero was struggling with his mask and was sobbing. 'What are we going to do now? Where have they taken Charlie?'

Mei-Ling put her arms around him and tried to comfort him. 'They may just have gone out for a sail, and will return. They'll return for sure. There isn't a better place to hide in Tequila.'

'Something must have tipped them off,' James ventured.

'Perhaps they heard us up on the cliff,'

'I don't think so. We were very quiet.'

'What are we going to do now, Mei-Ling?'

'We should enter the bay and have a look around,' Mei-Ling decided. 'They may have left some clue behind.'

Mei-Ling instructed the captain to ease the patrol boat through the passe. The powerful search light was turned on, and it illuminated the bay from one end to the other. The sea swallows started to agitate in their hollows, making some sorties to investigate the bright light. The patrol boat stopped in the middle of the bay, and continued beaming the light around the shore. There was a light signalling from the top of the cliffs. It had to be Maurizio and the police. Mei-Ling scoured the shore line with her night vision binoculars. She lingered over the decaying fisherman's shelter.

'I want to get to the shore, Captain. Can one of your men take me with the dinghy?'

'Yes, Ma'am, but be careful, there might be someone there.'

'That's exactly what I think. I'm sure I saw something move.'

'I'll come with you, Mei-Ling,' Sorciero said and she nodded. They climbed down into the dinghy and the coast guard started the outboard. They tied up to the jetty, and Sorciero jumped ashore. Mei-Ling shone her powerful torch into the fisherman's shelter. They could make out a curled-up figure cowering beside a post.

'This could be Charlie,' Sorciero bellowed and ran towards the figure, wielding his knife. The figure stood up and they saw that it was not a child, but a young woman with long blonde hair. She was tied to the post of the shelter and her face was contorted with rage.

'Untie me and take me away from here,' she snarled.

Sorciero stared at her in disbelief and then threw his head back and shouted: 'I'll be damned, if it isn't Maria Dolores.'

'I'll be damned if it isn't Sorciero, the oaf my ridiculous sister Maria Beatriz married,' she spat. 'Cmon, cut that rope, what are you waiting for.'

'Why don't you just keep a civil tongue in your big mouth, whore. No one will cut the rope until you tell us where Charlie is.'

'They took the little runt and left me behind, those bastards,' she raved. 'They got the wind up, because they thought they had been betrayed. They accused me of giving their hiding place away to the police. After all I did for that bastard Curria, he goes and leaves me here, to die of hunger and thirst.' She stamped her bare foot and squealed with pain as it hit a sharp pebble.

'Where did they go?' Mei-Ling asked quietly.

'How should I know, you stupid Chinese bitch. They just steamed away.' Tears of rage and frustration gathered in Maria Dolores' eyes.

'Well if you can't help us, we'll just leave you here to rot.'

'No, you can't do that; you'd become a murderer, if you left me here to die. Please help me,' she pleaded.

'We'll take her with us, Sam. Perhaps she'll remember some more when she's in the prison cell,' Mei-Ling said and threw Sorciero a pair of handcuffs.

'Don't you dare do that, you savage,' Maria Dolores yelled.

'So I'm a savage now... We'll just leave you here then to be rescued by a civilized person, eh Mei-Ling?'

'Good thought, Sam,' Mei-Ling said softly, then raised her voice. 'We're moving off, Captain.'

'OK damn you, put those cuffs on me, but you'll regret it, all of you.'

Chapter Nine

The following morning James drove to the Carlton Hotel and went in search of Muso Medusa. The bellboy, Rodrigo grinned from ear to ear, as James pressed a couple of tickets for the bullfight into his hand.

'So, where is Mr. Medusa this morning? Still having breakfast?'

'Oh no, Senor de la Cruz. He went very early this morning. Almost still the middle of the night. Gonzalez the night porter told me.'

'Out so early? Did he go for a jog?'

'No, Senor. He took his luggage; didn't want the night porter to help him. Probably too mean to give him tip. Anyway, the chauffeur from the limo helped him. A real slime ball, that Medusa.'

'Tut tut, that's no way to speak of a client, Rodrigo,' James admonished him. 'Where did he go? To the airport?'

Rodrigo shook his head. 'Told the driver to go to the harbour. That's all I heard.'

'Have you by any chance got the number of the limo company which took him to the harbour?'

'Yes, sir, if it was the same one that he used before, I just happened to notice that it was 3001 Te. It's the Luxury Car Co. They have several limos and all the number plates start with 300.'

James stopped at the US consulate and asked to see Jason immediately. He informed Jason of Medusa's departure, and the likelihood that he had been picked up by the yacht during the early hours of the morning. There hadn't been another word from the kidnappers and Maria Beatriz was about to have a nervous breakdown, according to Sorciero. To have been so close to success and been foiled by a whisker, was almost too hard to bear for Charlie's parents.

Maria Beatriz had been personally present when Maria Dolores had been grilled by Chief Pereira. She had admitted that she had joined forces with the Curria clan to seek revenge after her ignominious departure from Tequila. It seemed the natural thing to do. She had taken Romeo Curria as a lover when she had met him in New York at a social gathering. It was obviously the finger of fate that had brought them together. He was generous and she was fast running out of funds. Living in New York, in the style to which she had accustomed herself, was a pricey business. Romeo Curria had arrived in the nick of time. Together, they had made the plan to purloin Petrolia, and endow their respective nest-eggs, while at the same time inflicting a mortal blow to their common enemies. It had seemed a most delightful prospect, to hear Maria Beatriz begging abjectly for the life of her son, and to see the ruin of the Capponcinis and the Changs.

The first plan however hadn't quite worked out. Maria Dolores had had to abandon the abduction of Charlie, and Curria had moved in his henchmen to take care of the problem. Ever since then, Curria had as good as lost interest in Maria Dolores, and had started to slap her around, for no reason at all. Maria Dolores pointed to a bruise on her cheek, and some pressure marks on her ivory skin. She didn't elicit any sympathy from her gaolers, who looked ready to slap her around themselves. After repeated questioning and dire threats, they decided that she hadn't a clue as to where the yacht had sailed to, and that she had told them all she knew. The only important thing was that she swore that there was a mole at Police headquarters, who had informed Curria that a visit from the coast-guard was imminent. It had triggered the hasty departure of the 'Sharkstooth' carrying Charlie to some unknown destination.

Maria Beatriz and Sorciero sat disconsolately at the breakfast table and tried to swallow a morsel of food. They hardly spoke. Now and then Maria Beatriz wiped away a tear, and Sorciero held her hand and dabbed at her cheeks. They were waiting for a call from the kidnappers, and they hadn't decided what to do, or what to tell the kidnappers when they called. They were in a cleft stick, Petrolia on the one hand, and Charlie's life in the other. They would both give their lives to save Charlie's, but to give away Petrolia seemed an impossible gesture. To give away the life blood of Tequila, and plunge the island republic into abject poverty again was unthinkable. There was only one way, and that was to find Charlie before any deals were struck. They would have to string the kidnappers along somehow until the child was rescued and back home. The telephone rang and they stared at each other. Sorciero picked the cordless phone up with a shaking hand and answered. He looked surprised and put his hand over the mouth piece.

'It's the ole harridan from Mangorenia,' he whispered to Maria Beatriz. 'Yes Mère Magot, I can hear you.' Sorciero suddenly looked startled. 'You really did? And where was he? Hang on, I want to tell my wife.' He looked at Maria Beatriz and said: 'She's seen Charlie,'

'Where, for God's sake?' Maria Beatriz cried.

'In her crystal ball,' he grimaced.

'You can't be serious!'

'Sh...' Sorciero whispered and held his hand over the mouthpiece 'She can hear you if you shout like that.' Then he proceeded. 'OK, I's listening to you, Mère Magot, tell me more.' Sorciero listened intently to the hoarse voice of Mangorenia's Shaman.

'Who told you that Charlie's gone missing, Mère Magot.'

'No one at all. I just saw it in my crystal ball. The boy's on a yacht and the yacht is here, in Mangorenia. My crystal ball never lies, Sorciero. Your boy is in great danger, so you must do something at once.'

'Where in Mangorenia, Mère Magot?'

'That I don't know, Sorciero. Perhaps in the small harbour, or maybe the yacht's found its way into Dragon's Cove. I've heard say that there is a secret passe into Dragon's Cove.'

'I've never heard of a secret passe or of Dragon's Cove.'

'Only a very few people know about it; it is practically invisible from the sea, and a yacht might easily take cover there. No one ever visits that place; it is considered 'mal chance' by the Mangorenians. Come quickly, Sorciero.'

'If it's such a secret how come they were able to find Dragon's Cove? Tell me that, Mère Magot.'

'They must have friends in high places, so beware to whom you give your trust.'

Sorciero put the phone down and looked at his trembling hands. Who had betrayed them? That was the question. There were only a very few members of the old regime who had retained their positions after the cataclysmic upheaval in Tequila. As for family members of Big Daddy's, only Maria Beatriz had remained on the island, to become the President. Her three brothers had perished in the earthquake, and only her widowed sister-in-law had survived, and had married the Italian Consul. She was surely above reproach and had joyfully accepted her escape from a loveless marriage to an unfaithful and difficult husband.

'We must go to Mangorenia, Sweetness,' Sorciero said slowly. 'The Mère Magot is convinced that our boy is there. I must immediately get in touch with Jason and tell him about this phone call.'

'What about the kidnappers? Why haven't they phoned?'

'Perhaps their mobile doesn't work from there.'

'You must get Etienne on the phone, Lover. Ask him to discretely investigate the harbour, and find out if 'Sharkstooth' is anchored there.'

'My guess is that they're hiding out in that Dragon's Cove, but if they notice that a search has begun, they'll disappear again.'

'They may just be off shore somewhere; it's such a vast ocean,' Maria Beatriz wept.

'Don't cry, Sweetness; we'll get our Charlie back, safe and sound,' Sorciero tried to comfort her, feeling none too optimistic himself. After they had spoken to Etienne Bonheur, who vowed to do everything in his power to seek out 'Sharkstooth', and then had contacted Jason Reed at the US consulate and told him about the Mère Magot's phone-call. They hunkered down and waited for the kidnappers to make their move.

The long awaited phone call came sooner than they had anticipated. The line was not very clear, but the threats came over regardless. The kidnappers would wait only another twelve hours before sending the first severed limb of their young hostage to his parents. Exactly at 9 pm., the President, Maria Beatriz in person, was to meet the kidnappers, and an exchange would take place. The boy in exchange for the certified grant of Petrolia and all it's assets to a company named 'Tuttofrutto.' The venue was to be kept secret until the very last moment. Maria Beatriz accepted the terms and agreed to the meeting, but demanded to hear the voice of her son and know that he was alive and well. After a short silence Charlie's voice came over and after a brief conversation, the communication was broken off.

The Capponcini's speedboat threw up a trail of froth as it cut through the water. Miranda was at the helm and handled the boat like a professional. She stared straight ahead, not honouring James, who was standing beside her, with even a sideways glance. She had been persuaded to play her part in the game, and they were heading towards Petrolia, as James had been ordered to by Romeo Curria. They had obtained a landing permit from the Ministry of the Interior, and were going ashore, so that James could be seen to be obeying instructions. Miranda had collected him at the dock in Beatriz Bay and had turned her back on him without answering his greeting. A heavy silence hung between them, and it was a relief that as soon as they were out on the open sea, the noise of the engines and the crashing waves made all conversation impossible. They didn't appear to notice that another boat was following them, so engrossed were they in their respective dark thoughts. It took no more than ten minutes to reach the man-made dock which ran the whole length of the lee side of Petrolia. The atoll was a mass of industrial material. Pipe-lines, hoses, and holding tanks.

The government of Tequila had decided not to build a refinery, but rather send the crude to Venezuela to be processed there. It had been argued that there would be less chance of pollution of Tequila's pristine beaches. A deep water harbour had been created in Petrolia, and tankers were laden with the precious oil and sent across to Venezuela's refineries. Petrolia was heavily guarded by a special branch of the police force. A couple of coastguard cutters were constantly patrolling the area. James squinted at Miranda, but she didn't turn to him.

'Did you notice that we were being followed?' he ventured. She nodded.

'Do you know who it was?' Again she nodded.

'You do realize that it could be the mole, who's going to report back to Romeo Curria,' James said. 'What are you going to do about it?'

Finally she turned to face him, eyes stormy with contained anger. 'I'm certainly not going to tell you, traitor,' she said. 'I can't think why my parents and everyone else are inclined to trust you, now that they know all about you, and your mission impossible.'

'Dearest girl,' he started to protest, but she cut him short.

'Don't speak to me again, ever,' she said.

Jason Reed sat in his office and tried to keep his eyes off the pair of legs which sat crossed in front of him. They belonged to a woman whose exceptional good looks had turned several heads in her time. Not that she was past the sell by date now. He had seen her date of birth in her passport, and she was just a couple of years younger than he was. There was no great resemblance between her and her son, James Harvey De la Cruz. Only the clear green eyes, the ready, somewhat shy smile, and the regular features were common to them both. Her chestnut hair was well groomed and her complexion perfect. She was simply beautiful. An American picture-book rose. She had used the consulate guest suite, changed into a white linen suit and high heeled sling-backs, and now sat, seemingly cool and refreshed, opposite Jason. She couldn't however, hide her anxiety as she sat wringing her hands on her lap.

'Mr. Reed, is my son alright? Why wasn't he here to meet me?'

'Please call me Jason,'

'My name's Kate, but you know that already. So where is James?'

'He's doing a job for us, Kate, but he'll be back soon. Would you like a coffee or some tea? You must be exhausted after your journey.'

'I had some water in the suite. As for the journey, it was tiring, and also frightening. What is going on?' she demanded to know.

'James confessed the real reason for his presence on Tequila to us.'

Kate Harvey Curria tried to hide her reaction to this bit of information. She swallowed hard, before she whispered, 'So you know about our situation. I am so ashamed.'

'No need to be, Kate; I suppose that you did what you thought was right. He's a good man, your son. He quickly realized that he could never do what his father expected of him. James isn't a criminal.'

'They threatened us, Jason. I was afraid they would take James and murder him.' She gazed piteously at Jason. She looked so vulnerable and in need of protection, that he felt this great wave of feeling sweeping over him. He took her hand and squeezed it encouragingly. 'I understand, my dear; I have no children of my own, but I understand nevertheless.' Jason cleared his throat and stood up. He was in danger of completely losing his head over this almost unknown woman. He took off his glasses and cleaned them assiduously, so as not to have to look at her.

'I have been living in constant fear for years,' she continued.

'Why didn't you leave Curria and go back to the States?' he blurted out, hoping to hear a good reason.

'I was very young and frightened to take any action. I had a young child to bring up, and it seemed the right thing for James at the time. Besides, I wasn't immediately aware of the hornet's nest I had stumbled into. Romeo Curria was handsome and masterful. It felt good to be guided and loved by an older man. Afterwards it was too late, somehow. I got used to the solitary life on the ranch and James loved the farm and the animals. Besides, there was nothing to go home to. I was an orphan. I was given money to have an abortion, but I couldn't face that, so I took a job on a cruise liner to get away as far as possible from Boston.'

Jason looked nonplussed. He put his glasses back on and asked, 'You were pregnant before you met Curria?'

She nodded shamefacedly. 'I didn't know too much about contraception, or even sex. My father was a preacher and the family was very straight-laced. My parents both died in an accident. It happened during my last year in High School. I was very much in love with this boy, and thought he loved me too. We eloped and got married secretly in Las Vegas. His family managed to get an annulment and persuaded him to leave me.' Even now, her eyes filled with tears at the remembrance of her abandonment by the father of her child.

'Just let me get this straight,' Jason said slowly. 'You were already pregnant when you met Curria on board the cruise liner.'

'Six weeks,' she answered.

'And you got married.'

'It was like a sign from heaven, that a kind, handsome, obviously rich man would want to marry a stewardess. I jumped at the chance to give my unborn child a father.' Kate Harvey suddenly didn't seem so vulnerable any longer, Jason thought. It was just her misfortune that the husband was an already married man and a Mafia boss.

'So Romeo Curria never cottoned on to the fact that James was a cuckoo in his nest.'

She shook her head and stared down at her hands.

'And you never thought of contacting the biological father?'

Again she shook her head. 'I had made my bed, as the saying goes, and James was content.'

Jason started walking up and down the room, digesting the information Kate Harvey had volunteered.

'So James is under no obligation whatsoever to this Romeo Curria,' he said. 'I think you should tell James about the situation as soon as possible. It might change his life and the life of the girl he loves.'

'What girl?' Kate Harvey's mouth dropped open.

'Miranda Capponcini. The Capponcinis are some of the people James was supposed to destroy, here, in Tequila.'

The telephone rang on Jason's desk and he answered it. The conversation was pretty one-sided. He listened and just said 'yes' now and then. 'Come here immediately, Sam, we'll make sure we get Charlie this time.' Jason put the phone down only to have to answer it again. 'Yes, Miranda, but I can barely hear you.' He gasped when he finally got the gist of Miranda's message. 'O.K., I'll get the ball rolling,' he shouted and hung up. He looked at Kate Harvey and said, 'James will be here in an about half an hour, so tell him everything. Could you go back to the guest suite, Kate? I need to use this office for a meeting.'

'Certainly, Jason, but you'll keep what I've told you to yourself. Promise me that. I don't want James to hear it from anyone else.'

'You've my word, Kate.'

Sorciero and Maria Beatriz sat opposite Jason Reed and looked totally astonished as he told them that the mole in the local police force had been unearthed. It was none other than Chief Pereira, the trusted head of police. They shook their heads in disbelief at the news.

'Are you sure? How did you find out?'

'The Chief followed Miranda and James to Petrolia, but kept his distance. He obviously didn't want them to know he was on their heels. I'm sure that he was just checking out whether James was really going to investigate the oil rigs. Chief Pereira knew about the raid we were about to launch on 'Sharkstooth' while it was anchored at Little Bay the other night. He must have warned them of our plans.'

'Is Miranda quite sure that it was Chief Pereira?'

'There's no doubt that it was the Chief. Mei-Ling was out in her boat watching the whole thing, and she has just confirmed it.'

'But why would he do that? He has got top pay, a good house, a Range Rover, a nice wife and two kids. He even gets invited to the Government ball; what more would a fellow want?' Sorciero objected.

'I don't suppose it's the money,' Jason said thoughtfully. 'He could be blackmailed by the Mafia.'

'What about?'

'You never know... What if they took his wife or one of the children hostage, just like they took Charlie? That's a possibility, isn't it?'

'Supposing that was the case, where would they be keeping them?'

'All of this is just speculation. We'll have to question Chief Pereira. Ask him to come to the presidential villa for a briefing and we'll tackle him then.'

'OK, can I use your phone?'

'Sure thing, Sam. I'll dial the number for you.'

James saw the presidential limousine drive off, as he pulled up in front of the US consulate. It had been a thoroughly awful morning, and he almost felt sick to his stomach with mortification. Miranda had been true to her word and hadn't addressed another word to him all the time they had spent together. She had gone off into a corner of a shed and used her mobile phone, while he had moseyed around pretending to look at Petrolia's oil wells. The original one which had started to gush out of the small atoll after the monstrous earthquake, and several others which had been found subsequently. He was accompanied by a Tequila official who proudly showed off the whole island. James could hardly repress a series of yawns, and was greatly relieved when Miranda sent for him. She was already in the speedboat revving up the engine, and had taken off like greased lightening as soon as he had jumped in. He had fallen and hit his head on the command console, cutting a gash in his forehead. He had managed to sit back onto the seat, and find a handkerchief which he had pressed against the wound, to stem the blood. Miranda had indicated to him where the first aid box was located and he had taken out a plaster and applied it to his throbbing forehead. There hadn't been a word of sympathy or even a pitying glance from Miranda. He hadn't expected her to hold out so long. He was full of admiration for her steadfastness, but now began to wonder whether this was really the end of their romance. He still felt a little dizzy from the bump on his head, as he got out of the jeep and entered the consulate. The Marine eyed him, and called Jason's office announcing the arrival. Jason Reed looked at the young man as he came through the door, and walked towards him.

'What happened, James? Get into a fight with Miranda?'

'You might call it that,' James answered and put a hand to his head. 'I fell in the speedboat; truth is, I don't feel so good and my head aches, and I think I've broken a tooth.'

'You shouldn't have driven. Miranda shouldn't have let you take the wheel.'

'She doesn't care whether I live or die.'

'She does, my boy, she does,' Jason tried to console him. 'You do look a little pale. Come and sit down, and I'll get you a couple of aspirins. You might have a slight concussion.'

'Shit, that's all I need with the bullfight coming up.'

'We'll get a doctor to see you, and I'm sure you'll be alright for the corrida. I have a surprise for you that might make you feel better.' Jason called the Marine and told him to fetch the guest. James started up when he saw Kate Harvey Curria come through the door.

'Mother,' he cried, 'I'm so glad you're safe. Forgive me, Mother, but I couldn't do it. Could never do it,' he repeated and fainted at her feet.

When he came to, he was lying on a bed, with an icepack on his head. His mother was standing over him on one side, and a stranger on the other.

'What happened, Mother? Who's that? Are we prisoners of the Curria clan?'

'No, James we're not prisoners. This is Dr. Manners, who's going to examine you. You scared the life out of me, James,' she answered and held his hand. 'The doctor thinks you have a slight concussion, and have to rest for a couple of days.'

'What about the corrida?'

'He'll examine you again before the corrida, to see whether you're fit enough to fight; in the mean time, you have to rest. You'll stay here with me at the consulate, and I'll look after you.'

'Thank you, Mother.' He sighed. 'I have loads to tell you, about Miranda, and everything else.'

'I know, Darling; I'll have some things to tell you as well, but now let Dr. Manners examine you. Then you can close your eyes and rest.'

Sorciero and Maria Beatriz were sitting at the partners desk in her office. Old Crone Chang had made the herbal tea, and was handing out the cups. Maria Beatriz kept rearranging her flame coloured turban, and Sorciero was biting his nails. Tension was high, and the tea had a calming effect. They all looked up expectantly as they heard the front door bell ring. Chief Pereira bustled in, looking glum.

'I'm sorry to have to report that we haven't had any luck at all. There's no sign of 'Sharkstooth' anywhere. I took one of the coast guard cutters out this morning and scoured all the bays of Tequila without success.'

'I suppose you saw Miranda and the bullfighter going out to sea as well.'

'That I did; they were nosing around Petrolia, and even went ashore.' Chief Pereira looked puzzled. 'I didn't issue them with a permit, though.'

'No, I did,' Maria Beatriz said.

'That's alright then.'

'By the way, Chief, I meant to ask you, how is your family? We hope to see your wife at the Government Ball at the end of the week,' Maria Beatriz continued graciously.

Chief Pereira looked doubtful. 'She ain't too good these last few days. It's the flu. The kids have got it too. They won't leave the house for a while yet.'

'Flu, is it?' Old Crone Chang murmured. 'I've got just the medicine they need. I'll take it over right away.' She got up and started to pack her wicker basket.

'Nah, don't do that, Madame Chang,' Chief Pereira protested almost violently.

'I could come too, Mun,' Sorciero said enthusiastically. 'We'll have them fit and ready for the festivities.'

Chief Pereira squirmed on his chair, pure panic staring out of his eyes. 'You mustn't come to the house; I'll bring them over to your consulting rooms, Madame Chang.' He turned to Sorciero. 'Please, Sam, don't do it.'

Maria Beatriz stared at Chief Pereira. 'So they've got you by the short and curlies too, Chief Pereira,' she said. 'You might as well tell us everything, Chief, cos we can take a good guess at it anyways.'

Chief Pereira hung his head and stared at the floor. 'They came to the house and left a man there, with a gun. He threatened to knock them all off, one by one, if I didn't co-operate,' he said at last. 'Forgive me, Maria Beatriz, I didn't know what else to do.'

'You're one piece of shit, Mun,' Sorciero raved. 'It was your duty to tell us and we would have all helped each other.'

'Recriminations won't help any one now, Sorciero,' Old Crone said softly. 'We must make a plan to free all the captives.'

'We know that 'Sharkstooth' is hiding somewhere in Mangorenian waters. You're going to tell us where,' Sorciero butted in.

'How do you know that?' Chief Pereira stuttered.

'I'm not one of the most famous and powerful witch-doctors in the Caribbean for nothing, you asshole. Is it Dragon's Cove where they are hiding out?'

The Chief's round eyes became larger than blood-shot ping pong balls, as he stared at Sorciero in awe.

'Yea, Mun, that's where they are,' the Chief answered with a trembling voice.

Maria Beatriz grabbed the phone and dialled Jason

Reed's number. 'The Chief's confessed, 'Sharkstooth' is anchored in Dragon's Cove. Bring out the Marines, Jason, we're going to get our boy,' she shouted.

Kate Harvey Curria sat on the side of James' bed and put a fresh ice-bag on his head. He had slept for a couple of hours, and had just opened his eyes. He took hold of her hand and squeezed it.

'I'm so glad you're here and safe, Mother. It's been an awful couple of days.' She smiled at her son, and stroked his hand. 'I've got to tell you, Mother; I'm in love.' He smiled beatifically. 'She's the most beautiful girl you could ever see, and smart too.' He started to frown. 'She's not speaking to me at the moment, because of this whole Curria thing.' His eyes filled with tears. 'You must make her see reason, Mother. Explain it all to her.'

'Yes, Baby, I will, don't worry. Try to rest and not think about anything.'

'But I like to think about her,' he said and closed his eyes again. 'I want to think about her for ever,' he murmured and nodded off. She continued to stroke his hand until he was deeply asleep again. She got up and went to find Jason Reed. The Marine standing guard in the hall of the consulate, told her that the consul had left on urgent business, and wouldn't be back before dinner time. Dinner was usually served at 8 pm., if the consul ate at home. He would expect Mrs. Harvey to join him, if he returned on time. Otherwise, she would be served dinner in the guest suite, if that suited her.

Romeo Curria was sprawled out on a deckchair aboard the 'Sharkstooth'. He looked the epitome of a true yachtsman, wearing a T-shirt with the name of the boat embossed on it, and the captain's cap. He was sipping a Campari and soda, and was content. The yacht was safely anchored in the Dragon's Cove, and all he had to do now was wait for the Tequilans to deliver Petrolia to him. After all, he still held the trump card and he never for a moment doubted that his plan would succeed, in spite of the initial set back. Charlie boy was still securely cuffed to the bed rail in a stateroom below, eating chocolate cookies. The mole had delivered the warning and they had been able to escape from Little Bay just in time. Of course he had left Maria Dolores behind, and that hadn't been such a good move. They would no doubt grill her, but on the whole, there wasn't anything much she could reveal, except that they had had a relationship. She didn't know where the yacht was headed for, and where it would hide out. As for the rest, he didn't really care. Women came and went in his life, without leaving their mark. She, in any event, had outlived her usefulness.

The Dragon's Cove was eerily quiet, and completely hidden from view by a land-spit. One could quite easily have sailed past it and not realized that it was there, tucked away behind the cliffs. Furthermore the mole had assured him that it was never visited by the locals, because traditionally it was considered to be taboo by the denizens of Mangorenia. Romeo Curria idly wondered why it was so. It seemed an idyllic little cove, a perfect shelter from the brisk trade winds. The water was deep and dark blue and a small sandy beach looked tempting, but there wasn't a soul on it. A ragged path ran down the cliffs to it, which looked less tempting. It would be quite a rough climb to get back to the top.

The wind picked up as it usually did in the afternoon but the sun was still quite high and hot. It wouldn't take long though, for it to disappear behind the cliff. It was almost time to go fishing for some supper. Romeo Curria summoned up the cook, who stood cowering before him.

'Go and spear us some fish for supper, and be quick about it,' Romeo Curria commanded him. 'And stop feeding that child cookies. You'll make him sick.' The cook nodded and went to get his gear. Suddenly a great sigh rose from the cliff wall; it was like the lament of a woman in distress. It echoed around the cove, and struck fear into the hearts of the men on board. They gathered on the deck and looked to Romeo Curria for an explanation. Salvo Salsiccia, his second in command, cleared his throat and asked:

'What was that, Boss?'.

'Probably an echo of some kind,' Romeo Curria answered nonchalantly.

The wail was repeated, and they all looked around the cove, dreading to see something awful.

'The cook warned us about this place; he knows about it. They know about it all over the Caribbean. The local people never come here because of the "Malocchio",' Salvo Salsiccia stammered.

'That's precisely the reason we're here, Imbecile! We're safe in this place,' Romeo Curria glared at his crew. 'You are behaving like a bunch of hysterical women! Where are your coglioni? Huh?'

'Look, Boss, is there someone climbing down the path?'

'Here gimme the binoculars, I'll tell you what it is.'

That afternoon, the most daring rescue operation ever seen in the Caribbean took place. All the 'inner circle' of Tequila was involved, in one way or another. Maurizio Capponcini and two divers were let down with their diving gear from the sleek racing vessel of Young Chang, in the vicinity of Dragon's Cove. The divers entered the cove undetected and approached 'Sharkstooth'. A group of young people posing as foreign tourists had clambered down to the tiny beach in Dragon's Cove.

They had shouted greetings to the elegant yacht and proceeded to make a fire for a barbecue. They had a big wicker picnic basket, and a large cool box, which they started to unpack. Bottles of beer and wine appeared out of the cool box. Beneath the bottles nestled a couple of sten guns and ammunition. The wicker basket was emptied of the food, which had been hiding a couple of loaded pistols. Two of the men put on wetsuits and diving-gear, and surreptitiously tucked the hand guns into the suits. They picked up spear guns and were asked by the others to dive and spear fish for the barbecue. Amid a lot of ostensibly semi-drunken encouragement from the girls, they slid into the water and submerged. Then it all went very quickly. Divers surfaced all around the yacht and dealt quickly and cleanly with the armed guards aboard the yacht. There were no prisoners taken. Romeo Curria was shot as he raced for cover. They found Charlie handcuffed to a berth, alive and unharmed. He appeared to have weathered the storm remarkably well, and flew into his mother's arms as she ran towards him on the tiny beach.

Old Crone approached Chief Pereira's house on foot, carrying her basket. She tripped along in her tiny black Chinese slippers, seemingly unaware of any danger. She rang the bell and waited. A fat man in check shorts opened the door and glared at her.

'Whaddya want, old woman? There ain't no one here,' Muso Medusa growled.

'Oh... what shall I do with it then? The missus ordered a chicken from me for her dinner. It's in here.' She opened the basket, swiftly took out the silencer gun, and calmly shot the corpulent man in the chest. The look of amazement on his face was beautiful to behold as he slowly crumpled on to the floor. She stepped over him and made for what looked like the living room. She threw open the door and saw three terrified faces stare at her. The woman was the first to find her voice.

'What are you doing here, Old Crone? You must leave immediately. We are all in mortal danger here.'

'No more, you aren't,' Old Crone said softly.

'What happened to the guard?'

'I just shot him,' Old Crone replied. The children and Mrs Pereira gasped in horror.

'You actually shot the bastard dead, Old Crone?

'No, no, not quite dead. He's just sedated. They gave me a stun gun at the marine biology centre, and I tranquillized him. The police are just about to take him into custody.' Old Crone couldn't suppress a cackle. 'What a weird world we live in, don't we, Missus Pereira? Anyway, it's all over now, thank the Lord.'

Old Crone Chang sat in her rickshaw and watched the evening paseo on the waterfront. She rode down to the harbour and asked the rickshaw lad to stop and wait, while she took out a pair of mini binoculars and scanned the entrance to the harbour. She put her hand to her chest when she saw Young Chang's boat round the bluff and sail towards its berth. When she saw them all get off the boat, she breathed a sigh of relief. She got off the rickshaw and embraced her son and daughter-in-law; she held out her narrow little hand to Jason Reed and Maurizio Capponcini. Maurizio picked her up and twirled her around. She was light as a feather.

'Mr.Capponcini, please put me down. This is very unseemly for an old woman like me,'

'Nonsense, Old Crone, you're not an old woman. You'll live forever.'

'I assume that all went according to plan, Mr. Reed?' she asked.

'It sure did, Ma'am, and how about your little ambush?'

Old Crone smiled and said. 'It was a breeze, Mr.Reed, that's what they say isn't it? The element of surprise is a great opener for hostilities.'

Maurizio set her down on the ground and said. 'I think that we should have a little celebration. Why don't you all come to the Club for dinner. I'm sure Domitilla is dying to hear all about it.'

'Charlie was unharmed?' Old Crone asked.

'He was little bit shaken up, but he was comforted by his mother and father. He is a very brave child. They are staying over in Mangorenia with Etienne Bonheur for the night, and will return to the Island tomorrow for the Happening.'

Old Crone nodded. Things were beginning to get back to normal and she liked that. She was not a great one for agitation. She would allow herself to relax once the Semana Santa festivities were over, and she could concentrate on becoming a grand-mother. There was one more important thing she needed to know. She whispered into Jason Reed's ear, and he patted her hand. If the information had arrived he would bring it to the Club, he assured her.

It was another one of those impossibly picturesque sunsets which painted the Caribbean sky with a swathe of rainbow colours and gold tipped clouds. Miranda was sitting on the deck, glumly gazing at the display. Usually she sat there and drank in the splendour of the incoming twilight, but tonight she felt lost and alone in this bright universe. Her father had telephoned to let the women in his life know that the operation 'Rescue Charlie' had been a total success. He had also told them that he had invited the participants to a festive dinner at the Club. Miranda had immediately said that she wouldn't go if James were going to be present, and cried that she wished she'd never met James Harvey

Caramba De La Cruz. She was convinced that her life was completely ruined, over, kaput. Domitilla ached for her daughter, because she knew that Miranda was really suffering. When she was assured that James wasn't going to be invited, Miranda reluctantly let herself be persuaded to go and start dressing for dinner, and join the 'rescue party' in their celebration.

The party met in the reception of the Club Royale. There was still a sense of excitement and tension brought on by the rescue, and they all hugged each other, rather like a successful football team. Old Crone was almost crushed in the middle and begged for mercy. Jason Reed escorted her to the table in a secluded niche which had been reserved for them, and Maurizio poured the chilled champagne. There were a lot of questions asked and answered, congratulations extended, and gasps at the admirable courage which had been displayed by them all.

'And you know of course, the best part was Old Crone shooting Muso Medusa in the chest,' Jason Reed said with glee.

'You really killed the man?' Miranda stared at the fragile old Chinese woman in wonderment.

'No dear, just stunned him; like one would a rhinoceros, for scientific purposes,' Old Crone said modestly.

'Come to think of it, he does look as gross as a rhinoceros,' Miranda giggled. It was the first time she had laughed for couple of days. 'Where was this Muso Medusa anyway? I thought that he had left the island on the "Sharkstooth".'

'We were wrong about that. The Limo took him down to the harbour alright; they grilled the Limo driver and he freely admitted that he had driven Medusa down to the harbour, but didn't stay to watch where his client was going. Medusa didn't get onto a boat however, but went to Chief Pereira's house which is just around the corner from the harbour, and took over from the guy who was holding the Pereira family hostage. The mole, you see, was Chief Pereira, who was trying to protect his family.'

'Good God, no wonder our first plan misfired. Who would have imagined the Chief doing the dirty on us,' Maurizio said.

'It's understandable, Darling. Wouldn't you do the same to save your family?'

'I suppose so,' Maurizio agreed reluctantly.

'What is going to happen to the Chief?' Domitilla asked.

'I expect that Sam is going to put a hex on him,' Mei-Ling said, 'and then Maria Beatriz will pardon him. He was truly in deep shit.'

'And Romeo Curria?' Miranda wanted to know.

'He, I'm afraid, is history. Well and truly dead, with a burial at sea.'

'I wonder what James is going to feel about all that.'

'At the moment he will feel nothing but a big pain in the head, when he wakes up. You should know that, Randy; he fell in your speedboat. He has had a concussion, and is under sedation,' Jason said and looked to see how Miranda would take the news. She had the grace to look slightly guilty.

'Where is he? Here, in his cottage?'

'No, he is at the consulate in the guest suite.'

'All alone?' Miranda's pity was obviously aroused.

'No... he has some feminine company to look after him,'

'Oh, has he now; I should have taken more notice when "OLA" branded him as a womaniser!' Miranda tossed her mane in contempt.

'You mustn't always jump to conclusions, dear girl. Mrs Harvey, his mother, has arrived and is looking after her son. She's a very good-looking, cool woman.' Jason couldn't keep the admiration out of his voice.

'So you fancy the lady, Jason?'

Jason blushed and looked embarrassed. 'Ridiculous, Maurizio,' he growled.

'Will James be able to take part in the bullfighting?' Mei-Ling asked, diverting everyone's attention.

'Dr. Merchant said that after a couple of days rest, he would probably be fit to do so,' Jason replied.

'Now the big question is, have we got rid of all the bad guys?' Old Crone threw in.

'I believe we have. Muso Medusa is behind bars, where he will be joined by Ms. Maria Dolores Del-Rey. As for the crew and passengers of the "Sharkstooth," they all perished at sea.'

'What happened to the yacht?' Domitilla asked.

'It too was lost at sea, after some sort of explosion. The report states that it was probably the butane gas which was used for the barbecue.'

'E se non e vero, e ben trovato,' Maurizio murmured.

'And what does that mean?' Young Chang asked.

'It's an Italian saying, which translates roughly like this: If it's not the truth, at least it's a good story. It's what has been logged by the coast guard and that's the end of it.' Maurizio looked around at his guests and they all smiled knowingly at each other.

'I've ordered dinner for us. I hope that you all approve. A pasta with asparagus tips, and grilled prawns accompanied by fried courgette thins.' There was a hum of approval from everybody. Jason bent down and opened the briefcase he had brought along.

'One more thing, before I really start on the champagne. I've got the lab results you wanted, Old Crone. Care to look at them now, or later?'

Old Crone held out her hand in anticipation. 'Now, Jason, please. It could be very important.' Old Crone opened the envelope and looked at the chart which listed the properties of the spring water in

Mangorenia. She started to smile and then to laugh. It wasn't a thing she often did, and it left the others wondering. Their curiosity was whetted, but Old Crone kept them in suspense.

'Well, Mother, are you going to let us in on the joke?' Young Chang finally asked.

'My son, it is so unbelievable that I can hardly believe it. Sorciero might not have found oil on Mangorenia, but we've discovered something else there, almost as profitable perhaps.'

'Don't tell us that it's all in the water you brought back from Mangorenia, Mother?'

'Exactly that, my daughter.' Again she chuckled and clapped her hands together. 'Who will hazard a guess?'

'Let me guess first,' Maurizio said. 'It contains the elixir of life.'

'That would have made me faint not laugh, Mr Capponcini.'

'It's a good cure for liver diseases,' Domitilla ventured. Like most Italians, she was always preoccupied with that organ in her body.

'That's surely not original enough. There are lots of spas in Europe that specialize in that area,' Jason said. 'I think it's a cure for loss of hair.' He passed a palm over his thinning crown. Old Crone shook her head and said: 'That's quite a good thought, but not the right one.'

'It must be sexual, then,' Miranda speculated.

'It must be, dear girl, and it is,' Old Crone almost crowed. 'The spring water which gushed up in Mangorenia after the earthquake, contains one of the most potent aphrodisiacs known to man.'

'Are you sure, Old Crone?'

'Totally sure. We have known about it in China for ever, but it comes from a tiny very rare plant and is very difficult to produce. One needs to distil whole acres of those little mauve flowers in order to obtain a small amount of the substance. It commands huge sums of money, more than the poor rhinoceros' horn.'

'What is this weed called?' Jason wanted to know.

'We know it as the flora erectus.' Old Crone whispered.

'As opposed to the fauna erectus, I presume,' Jason grinned.

'Well, whatever you like to call it, Consul; I saw a field of these little mauve darlings grow near where the spring is located. That, together with the, how shall I put it, the strange behaviour of the people who bathed in the natural spring pool and drank the water, put me in mind of flora erectus.'

'Are they bottling this curious liquid in Mangorenia?' Maurizio asked.

'After a fashion, they are. They've got an antique crown cork bottling system and use any old bottles, mostly Perrier ones. The water tastes slightly odd, not bad but different.'

'Did it have any effect on you, Mother?' Young Chang asked.

'Never forget to respect your parents, Young Chang,' Old Crone scolded her son, but her old eyes twinkled.

'Are you sure of your diagnosis, Mother?'

'The properties are there in black and white.' She waved the document. 'Besides, every female on Mangorenia seems to be with child, and that includes the animals.'

The group sat silently now, and pondered the revelation of the presence of 'flora erectus' in their immediate vicinity.

'So not only does it have potent aphrodisiac properties, but it could possibly also be used as a fertility treatment,' Domitilla broadened the spectrum of the potential of 'flora erectus'.

'Indeed, it really looks like a good possibility.'

'Do you think it is more effective than Viagra, and will it be able to compete with the drug?' Maurizio asked.

'Of course there will have to be some trials and research, but I can quite see Mangorenia becoming a Mecca for people seeking satisfaction,' Old Crone murmured. 'After all, what could be more pleasant than to spend a week or two in the Caribbean, in the lap of luxury, and be able to continue the good work at home by drinking the water.'

'We could open a Club Royale Spa there; what do you think of that?' Miranda said excitedly.

'Whoa, hold your horses, young miss, I wouldn't want you to work there,' was Domitilla's comment.

'Don't worry, Mamma. I won't drink the water, I promise.' Miranda grimaced.

'Sorciero and Maria Beatriz will be absolutely amazed at this development.'

'Probably so, Consul; they're due back tomorrow and Sorciero and I have to prepare for the Happening.'

'Tell them to bring some cases of water with them. It might make things happen at the Happening,' Jason grinned in anticipation.

A steaming platter of pasta was brought to the table, and presented to them. The head waiter deftly filled all the plates, and milled some fresh parmesan and pepper over them. The guests all beamed at each other and inhaled the delicate odour of the green asparagus, and the pungent bite of Grana cheese. They started to relax after the great strain they had been under, and instinctively held out their hands to each other, around the table. They knew that each one of them was thanking God for the delivery of Charlie and the safe return of the rescuers. This communion lasted a few seconds, and then they started to eat in silence, savouring the food and nodding their appreciation from time to time.

Chapter Ten

Etienne Bonheur presided over the dinner table, elegant as usual, in impeccably tailored trousers, and a silk three button polo shirt. An Hermes cravat was loosely tied at his neck. His wife Marthe was equally elegant in her latest Missoni knit. Etienne was most affable towards his guests from Tequila, and he was reaping thanks from both Sorciero and Maria Beatriz for the help the minute Mangorenian police force had contributed to the rescue of their beloved Charlie. In truth, Sorciero never was quite comfortable in the presence of Etienne and Marthe. He felt that he had to watch his language, his manners and his dress. Maria Beatriz had tamed him and tidied him up; had taught him the rudiments of behaviour and brought out the best in him. Nevertheless Sorciero couldn't shake this feeling of inferiority which crept over him when he was in the Bonheurs presence. Sorciero had got accustomed to living in the luxury of the presidential villa, but it was in his mother's old house in the woods that he felt most at home. There he could be completely himself could feel the power with which he was endowed, could commune with the spirits of his ancestors, and relax in the soothing presence of Old Crone.

Both Sorciero and Maria Beatriz knew that they had to add something tangible to the thanks they had extended to Etienne for his help in getting Charlie back. They had discussed this in their suite in the presidential residence, while Charlie was in the nursery, playing with the Bonheur children. It would have to be in the form of yet another substantial, interest free loan. Maria Beatriz stood in front of the mirror and rubbed coconut oil into her body. Sorciero came up behind her and grinned wickedly.

'What have we here?' he cooed and squeezed her purple nipples. She put her hand behind her and echoed, 'What have we here' as she squeezed his erection. Their coupling seemed to last for an eternity, and they both had to stand under the shower again and were late for dinner. Being late for dinner was frowned upon by Marthe Bonheur, as her French chef insisted on preparing a souffle of some kind for first course.

The atmosphere was electric around the dinner table. The aftershock of the afternoon's happenings was still pumping the adrenaline. At Sorciero's request, the Mère Magot had been invited to the dinner, and she had arrived adorned in her traditional all white shaman dress, clutching her crystal ball in one hand, and Cocorico, her miniature rooster in his cage in the other. Even though the Bonheurs were slightly in awe of her, Marthe drew the line at having the rooster sit at table with them, and it was banished to the pantry, where it complained loudly. It was eventually pacified by one of the kitchen maids, who fed him some sweet corn. Mère Magot placed the crystal ball in front of her and glared at the company. Her deep black eyes stopped at Sorciero, and then she smiled with all the brightness of her new dentures, which she only wore on special occasions. She rearranged her white turban and addressed Sorciero.

'Tell us of your exploits again, great Sorciero,' she commanded and Sorciero was made to tell the whole story of Charlie's rescue again, to the accompaniment of the ohs and ahs of the others. Sorciero and Maria Beatriz held hands across the table, and shed a few tears. Plans were made for their departure in the morning, and Maria Beatriz offered to send the chopper back to bring the Bonheurs to Tequila for the rest of the pre Semana Santa celebrations. There was Sorciero's Happening, the bullfight, the celebration mass at the Cathedral on Palm Sunday morning, and the Government Ball as the closing event on Sunday night. Suddenly Mère Magot put down the pork rib she had been gnawing on, and commanded everyone to hush up. She stared at her crystal ball in agitation and waved her hands over it.

'One of them got away, Sorciero!' she cried. 'He saved himself; I can see him runnin in the fields. He is startin to climb the side of the mountain, towards the spring. Maybe he's planning somethin bad. 'Et puis merde... perhaps he's going to poison our water to punish us for helping you save Charlie.'

'Etienne, call out the police,' Marthe urged, and everybody started up from the dinner table in horror. The cheese souffle was left to deflate and languish on the gold rimmed plates. Etienne called his Chief of Police, instructing him to bring as many armed men along as he could muster. They would meet at the bottling plant, which of course was shut down for the night. There was no night shift and no guard. They all went outside and waited for Mark to bring the people carrier around. They all piled into the van, and Mark started the engine. The kitchen maid ran after them carrying the bantam in its cage. It was handed in through the window, and rested on Mère Magot's lap, together with her crystal ball. Mark drove down the road, through Mère Magot's village, and stopped the van at her garden gate. She insisted, however, on going all the way with them and Mark drove up the slope towards the spring. The moon was high and the night not as dark as it might have been. One small, low-voltage lamp glowed outside the bottling plant building, and the place was deserted. Sorciero and Mark got out of the van and inspected the lock on the door. It hadn't been tampered with. The police van lumbered up the road and parked behind them. Four constables and the Chief clambered out, toting sten-guns.

'Did you see anything, Chief?' Etienne asked.

'Nothing, Mr. le President, but there are lots of hiding places up here, so we had better go and have a look. I'll leave two men to guard the building here, and take two men up to the pool, and see what gives.'

The Chief and his men walked cautiously up the path towards the spring trying to keep as quiet as possible. One of them was carrying a powerful torch, which lit up the path, and the trees at the side. It rustled here and there in the bushes, but that might have been insects, squirrels or birds. There was no sign of a human anywhere, and they reached the pool without interference. They shone the torch over the water and into the bushes overhanging it. The moon provided the rest of the light.

'I don't think anyone's here,' the Chief whispered.

'Mère Magot's usually right in her predictions, Chief,' his second in command whispered back.

'Let's just take cover here awhile then,' the Chief replied and they crept behind the hibiscus bushes and settled down to wait.

'This place is spooky at night, Chief,' the second-in-command whispered.

'Rubbish, mon brave, what's there to be spooky?'

'Perhaps it's the moonlight, Chief, but I'm sure we're going to see a ghost, an evil spirit.'

'Nonsense, there's nothing here. Not yet anyway.'

'Look over there, Chief, can you see that?'

'What do you see?'

'A figure in white. All in white. Mon Dieu, it's moving towards us. What are we going to do?'

'You're going to challenge it, that's what.'

'Why me, Chief? I've got a wife and three kids, with another on the way.'

'So, don't fuck so much is my advice.' The chief all but pushed his second-in-command out onto the path. The man pointed his gun at the figure and said with a trembling voice, 'Put your hands up, Evil Spirit or I'll shoot.'

'Espèce d'idiot,' growled Mère Magot. 'You should point your gun in the other direction. He's hiding under the water; go get him.'

'But he can't be under the water for that long. How is he breathing?'

'With some kind of tube, or reed.'

'Like in the movies?'

'Do I look like I go to the movies, imbecile? I saw it in my crystal ball.'

The men fanned around the pool and examined it more closely. The Chief suddenly pointed to a rubber tube amongst the reeds. It blended in well with the vegetation. They shone the torch directly on to the tube, and sure enough they could now distinguish a body below the surface of the water.

'Go get him, he's the evil spirit,' Mère Magot cried, urging them on.

'I can't swim,' the second-in-command objected and stood back.

'It's not deep, so stop arguing and get in there,' the Chief shouted at him. Reluctantly the men waded in, lead by the Chief himself. Suddenly the body hiding under the water leapt up and shook himself like a dog. He gnashed his teeth like dog and almost barked like a dog.

'OK, don't shoot, I'm giving myself up,' the man yelped.

'Who are you?' the Chief asked.

'I was the cook on board "Sharkstooth". I was out spearing for fish in the cove, when it all happened. I swear I don't have anything to do with the clients. We were chartered out of Curacao, you can check up on that. I's been the cook for three years now. Never had no trouble like this before. A bit of drug running, maybe some arms smuggling, but never this kidnapping shit.'

'What about the rest of the crew? Where they the regular crew of the "Sharkstooth?"'

'No, sir. The clients insisted on taking their own crew. They couldn't find a replacement for me, so I had to go along.'

'Didn't the owners find that odd?'

'Quien sabe? The clients paid a lot more money than other guests.'

'So what you doing up in these parts?'

'I swam ashore and was hiding. I didn't want no trouble with the coast guards, mun.'

'So why hide up here; why not go to the harbour and try and hitch a ride like?'

'I dunno where the harbour is, so I just walked and hid and walked and hid until I came up here.'

'This just sounds like a whole lot of lies to me,' the Chief said. 'You knew all the time what was going on, didn't you?'

'No sir, not me. I swear I didn't. When I saw what was happening, I tried to help the little boy. Brought him some coke and chocolate biscuits. He was damn near famished. You ask the little boy, he'll tell ya.'

'Any one else got away? Anyone else hiding here?'

'No sir, no one but me.'

'Put the cuffs on him, and take him down to the station. We need to check out his story,' the Chief said. 'What do you think? Anyone else lurking around here?' He looked at Mère Magot questioningly.

She shook her head. 'I only saw one.'

'Then we can all go to bed at last,' the Chief muttered, and led the way back to the vehicles.

The next day, the chopper brought Sorciero, Maria Beatriz, Charlie and the Mère Magot back to Tequila. Before their departure, the 'Sharkstooth' cook was brought to the Presidential residence, and was confronted with Charlie. The boy looked at the cook and smiled with anticipation.

'Have you brought me some chocolate cookies again,' he asked and held out his little hand.

'No more chocolate cookies today, Charlie, they'll ruin your teeth,' Maria Beatriz admonished, and led him away to the waiting car. The cook wept with relief and demanded to be released, and given the means to return to Curacao. His request was granted, albeit reluctantly, since there seemed to be no reason to detain him. He was taken down to the police station, where a bed was put up in a cell for him. It would be his abode until he would board the next container ship which went Curacao way.

Tequila was festively adorned, as it always was in the days before the Semana Santa. People were out on the streets, in their most colourful attire. Balloons were constantly released by kids who watched them fly high in the sky. Calypso music filled the air, and the young women swirled around in their flamenco style, many layered skirts. Delicious, spicy aromas were already emanating from the barbecue stands. The ubiquitous sausages were browning nicely and then were laid out on banana leaves, and garnished with fresh pineapple. The older denizens were walking around waving palm fronds, and gossiping in the Cathedral. The Cathedral was adorned with massed bouquets of white orchids, and sweet smelling white lilies. The aisles were festooned with white satin ribbons and bows in preparation for the Palm Sunday mass. The choir was busy practicing their hymns which had been slightly calypsoed up by the choir master, to the delight of the choir boys, and the disapproval of the old bishop.

The people on the street cheered when they saw the presidential family leave the airport in their Range Rover. Young Charlie rolled open the window and waved to them. The word had got about concerning his abduction and subsequent liberation, and they surged into the road, blocking the vehicle for a while, grabbing at his little hand and chanting his name. At last the family was allowed to proceed and drove back to the presidential villa. There were more welcoming celebrations at the villa. The old dog Poppy raced around them frenetically, until she collapsed at their feet from exhaustion, and all the staff where aligned outside the front door, waving palm fronds. Charlie was passed from arm to arm, kissed and hugged and petted. It was a homecoming none of them would ever forget.

James Harvey de la Cruz opened his eyes and stared at the ceiling. An old-fashioned cut-glass chandelier was hanging there, reflecting the sunlight in each of its prisms. It was an unfamiliar ceiling and he wondered where he was. He turned on his side and winced. He touched the side of his head and felt the bump. Now he remembered that he had fallen in the speedboat, and that he had felt pretty rough after a while. His head had hurt like hell, and his eyes had played tricks on him. He remembered driving to the consulate and that his mother was there, miraculously holding his hand and putting an icepack on his bump. After that he remembered nothing. Perhaps he had dreamt it all. He looked at his wrist-watch and whistled through his teeth. It was midday and he was still lazing around in bed. What made it worse is that he didn't know whose bed it was. He sat up and yawned. His headache had gone; his eyes were clear and his sight sharp, but he felt a strange ache in his stomach. His nostrils flared and he thought that he could smell fresh toast. His mouth started to water and he realized that the ache in his stomach was an angry hunger pain.

The door opened and his mother looked in. She smiled at him, and carried a tray to the bed, which she balanced on his knees. She bent down to kiss him.

'So it wasn't a dream after all,' he said. 'You're really here. I'm sure glad to see you, Mother; I was worried about what they would do to you if I didn't follow Curria's orders.'

'It's alright, Darling,' she comforted him.

'Nothing's really alright, Mother. You see I can't do what he wants me to do. I just can't.' he murmured.

'I know that now, Darling; I should never have asked you to do it. Fear is a terrible thing. I was afraid for you and I was too much of a coward to cross Romeo Curria.' She let her hand glide over his brow. 'Stop worrying about it.'

'I can't stop worrying about it,' he murmured. 'I've fallen in love with the girl whose family I was supposed to wipe out or something. Now that they've found out who I really am, and why I came to Tequila, Miranda won't even speak to me and I feel suicidal.'

'You can stop feeling suicidal, Darling, because I have something to tell you which will change your life.'

'Yea? Like what?'

'Like Romeo Curria isn't your father, that's what.'

'It's kind of you to try and comfort me with a fairy story, but this is real life, Mother,' James answered acidly.

She smiled at him indulgently, and continued to stroke his forehead.

'It's no fairy story. I was pregnant when I met Romeo Curria on the cruise boat. Pregnant, alone and desperate. I had been abandoned by your father, and had only distant relatives. It seemed like a heaven-sent opportunity to find a husband who was unattached, handsome and obviously rich. At least I hoped to secure your future.'

'You can't be serious! Secure my future by delivering me into the eye of the Mafia hurricane?'

'I didn't know, James. I truly didn't know until it was too late.'

'We could have left the ranch and gone back to the States. They wouldn't have found us.'

'Oh yes, they would, if they wanted to. Romeo Curria never doubted that he was your father, and he wouldn't have relinquished you that easily.'

James picked up the buttered toast and wolfed it down. He drank the coffee and tried to make head or tail of this confession.

'So I am under no obligation to kill anyone? Romeo Curria is not my father?' He took another piece of toast and chewed on it. 'Then who the devil is my father?' he asked.

'His name is Bruce Winthrop, the fourth. The family comes from Boston and goes back a long way.'

'Is that all you're going to tell me about him?'

'We were high school sweethearts, and decided to get married, when my parents died in a car crash, and I was told I was going to go and live with some relatives I had never heard of. There was no money and no alternative. So we eloped and got married in Las Vegas. His parents traced us and separated us. They managed to annul the marriage as we were both minors. We were forbidden to see each other and I was given $1000 dollars and a ticket to ride to Pittsburgh where my relatives lived. Instead I took a job as a stewardess on a cruise boat, which left for Argentina. I met Romeo Curria, who was a first-class passenger. He had the best suite on the boat, and I used to turn down his bed every night.'

'You not only turned it down, but got into it as well,' James said wearily.

She frowned at her son. 'I had been sick a lot, thinking that it was seasickness, but the ships doctor put me wise,' she said. 'I never got into his bed before the wedding ceremony which the captain performed just before we landed in Buenos Aires.' She bowed her head. 'I thought that it was all legal, but I found out later that Romeo Curria had a wife in Sicily. She was alive and well, not dead and buried as he had told me. He soon tired of me, but was very pleased to have a son.' She lifted her head, and her eyes glistened with tears. 'You were the most beautiful baby.'

James looked at his mother and thought how lovely she still was, and what a wasted life she had led on the ranch in splendid isolation.

'How could you shut yourself off from the world like that, Mother? Weren't you lonely and miserable, in a strange country?'

She shook her head. 'Lonely sometimes, but never miserable. I had you, James, and that was all that mattered.' She took the tray off his knees, and put it on the dressing table.

'One more thing, Mother. Did you ever tell this Bruce Winthrop about me?'

'No... there seemed to be no point. His family would have just branded me a gold digger, and disputed the paternity. Besides, he probably married that little snooty bitch Constance de Bier. That's what his parents were wanted him to do.'

'Gold digger? Were they that wealthy?'

'That they were, and then some. Have you never

heard of Winthrop Chemicals?' James shook his head.

'Well, it's all water under the bridge now, Darling. You've grown into a very fine young man without their help.'

'I still think that my father ought to be told,' James mused. 'You see Mom, I want to be able to tell my girl who my real father is. That he is an honourable man, not a Mafioso'

'Alright then, James; I can try and trace him'

They sat in companionable silence for a while, and James lay back on the pillow and closed his eyes. Kate Harvey rose and picked up the tray. He reached out for her, and grabbed her skirt.

'Don't go yet, Mom. Tell me more about my father.'

Kate Harvey put the tray down, and sat on the side of the bed again.

'He was the sweetest guy and we were so in love, but we were only seventeen when we eloped. If his parents had approved of our marriage, I suppose we would still be together. He was tall and blonde, like a Viking. A football player on the high school team. A real lion on the field, but so gentle and loving with me. I'll never forget how we cried, when they found us in Las Vegas and tore us apart. He swore he would come for me when he had graduated from college, but I didn't want to be a burden to him, so I took the job on the cruise liner.' She shook her head ruefully. 'Even if he had tried to find me, it would have been an impossible task. I had disappeared off the face of the earth.'

'Mmm... maybe he did try. I'll give him the benefit of the doubt.' James smiled happily. 'Tell me, Mother, is it really gone midday?'

'Yes, James. You've been asleep for more that twenty-four hours.'

James started up and stared at his mother. 'twenty-four hours? Good God, it must be Saturday then. I've got to get up and look at the bulls again before the corrida. And there's the Happening tonight. I promised Miranda that we would go together.'

'I thought that she wasn't speaking to you,' Kate Harvey said.

'You're right; that little detail slipped my mind. Oh shit, just plain good old-fashioned shit,' he moaned.

'Really, James, there's no need for such language. You are just going to be good and patient, so that you'll be alright for the bullfight tomorrow.'

'Do you mean that I'll have to hang in here until then? And anyway, where the devil are we, and whose bed am I sleeping in?'

'In mine, Darling, I've got a camp-bed in the sitting room. This is the guest suite in the consulate. Jason Reed is our kind host.'

'Jason Reed,' James said thoughtfully. 'He could find out about Bruce Winthrop, couldn't he?'

Kate Harvey inclined her head and smiled wryly. 'I daresay he could. I'll ask him to do that.'

'If we can find a telephone number, we could phone him, Mother.'

'I don't think that I could speak to him, James, not after all this time. Besides it's the week before Easter and many people will be on holiday.'

'I wonder where he is,' James speculated.

'Most likely still in Baltimore, where the company is located. He must be working for his father. That was always his plan; to become a research scientist.' Kate Harvey looked at her son who was the image of his father; perhaps it wasn't too late after all, to reunite them. James suddenly sat bolt upright, a look of horror on his face. He caught his mother's hand and squeezed it.

'God, I've been so busy with myself that I forgot all

about Charlie. Did they manage to save him?'

'Yes, James, the boy was freed and is alive and well. It was a combined effort of the President's husband, the Capponcinis, young Mrs.Chang, her husband, his mother who they call Old Crone Chang, and the Mangorenian police. Romeo Curria's yacht had taken refuge in a small cove in Mangorenia. It's a confusing story, but the boy is safe; the yacht sank, and the crew and Romeo Curria were lost at sea.'

'You mean Romeo Curria is gone for good?' James couldn't keep the satisfaction out of his voice.

'So it would seem, God rest his soul.'

'C'mon, Mother, I hope he doesn't.' James swung his legs over the bed, and got up. 'This is cause for a celebration, Mother.' He swung her off the bed and started to whirl her around the room. A knock on the door interrupted their dance. Jason Reed opened the door and looked in.

'I hope I'm not disturbing; may I come in?'

'I'm celebrating with my mother, as you can see.'

'I do see... that you've as good as recovered from your concussion, which is encouraging news. I'd like to take your mother to the Happening tonight, but she didn't want to leave you.' Jason Reed turned to Kate Harvey: 'Well, are we on, Kate?'

'I suppose we are, Jason,' she smiled at him.

'Then I'll pick you up at six o'clock sharp,' he strilled back and left the room.

'Mother, you're flirting with him!' James said sternly.

'Nonsense, James, of course I'm not.' She squinted at him. 'And even if I did, what's wrong with that?'

'What about my Dad? Perhaps he's been waiting all these years for you to contact him, and what are you doing? You're making eyes at the first man who comes your way.'

'Now don't go daydreaming that I and your Dad are

going to get together again. That only happens in the movies, Darling.'

'It's the kind of movie I like, Mom. Can't we make it come true? Life is stranger than fiction I've heard it said.'

'Would you like some grilled fish for lunch?'

'What's grilled fish got to do with anything? Don't try and change the subject, Mom.'

'I think you're getting over-excited. I think that you should lie down again and have another rest. Think of the bullfight.'

He squared his shoulders and stood his ground, staring at her.

'Save that look for the bull, Darling. Now go and brush your teeth and wash your face and get back into bed. I shall be back with lunch presently.'

Miranda was getting ready for the Happening, even though her escort had fallen by the way side. She was going to the Happening for the first time, and that in the company of her parents. It was almost shameful. Every girl on the island went to the Happening with their beau, because strange things were likely to occur at the Happening. Couples got engaged, men left their wives, children were conceived, and women found themselves waking up next to strange bed-fellows the next morning. Only Miranda was going to be left out of the fun. She brushed out her hair in front of the mirror and made a nasty grimace. Her young heart was aching, fit to breaking in her chest. She smoothed down her clinging silk mini dress and thought of James. James, who lay injured in a bed, and probably was missing her and feeling as miserable as she was. She felt guilt rise like bile in her throat. It was all her fault that he was afflicted with this nasty concussion. She had deliberately started the speedboat while he was still standing, and had laughed out loud when he had lost his balance, fallen and cracked his head. She had been gloating over his discomfort, and hoped that it would hurt like hell. Now she was devastated at her behaviour. She had heard from Jason Reed about the arrival of James' mother, and was curious to meet her, and see what she was like. Miranda took another look in the mirror and then turned away. What did it matter what she looked like if James wasn't going to be there.

The sun had dipped into the ocean, and people were streaming out of their homes and wandering down the road in the twilight. Rickshaws were transporting the more fortunate ones to the Happening. Cars were not allowed on the roads in the vicinity of the Happening. The sky was clear and the stars were starting to emerge. The Capponcinis left their Range Rover at the edge of the rain-forest and wandered towards Sorciero's ancestral cottage. A path of burning coals was glowing in front of the cottage in preparation of Sorciero's annual ritual. Every year he fell into a trance and promenaded down the path of searing embers in his bare feet. The crowd cheered him on and danced themselves into a frenzy.

Tonight Sorciero and Old Crone had a guest, la Mère Magot, who they installed in the Mandarin chair which Old Crone usually occupied. La Mère Magot had unearthed a curious outfit from her sea-man's chest. It had been left behind by a member of a Hungarian folklore dancing troupe, in lieu of payment for services rendered by the Mère Magot. She wore red leather boots on her bare feet, a Hungarian multi coloured layered skirt, a Hungarian peasant blouse and the classic, beaded headdress. She looked for all the world as though she was about to do the Czardas. The miniature bantam rooster, Cocorico, sat on her shoulder and pecked at the beads on her headdress. In her right hand she held her precious crystal ball. Old Crone Chang wore her heavily embroidered Mandarin coat and sat beside the Shaman of Mangorenia on a bar stool which she had borrowed from the Club Royale. Sorciero stood on the other side of the Mère Magot, sporting a leopard loincloth, and a fake fur leopard coat draped over one shoulder which he had purchased from Saks Fifth Avenue in New York. The ancestral feathered head-dress which was traditionally handed down from father to son, had had a make-over, and was glossier and fuller than before. They were an awesome, if bizarre trio, which presided over the Happening, and the onlookers were agog with excitement waiting for the surprise which was in store for them this year. Two rows of chairs were set up in front of the fiery path. They were reserved for the Government Ministers and foreign Diplomats.

By now it was completely dark, and the flambeaux were lit around the house and the platform where the trio sat. The throng of onlookers had become enormous, and there was some jostling for position. The Cabinet ministers and other dignitaries had taken their seats. Maria Beatriz sat in the front row wearing a gold thread pareo and satin top, with matching turban. Massive gold earrings dangled from her ears. The President of Mangorenia and his wife sat on either side of her, elegant in the newest Paris fashions. Beakers of cool water laced with syrup were handed round free of charge, and Old Crone had the hint of a smile on her lips as she watched the people drink. Jason Reed with Kate Harvey on his arm stood somewhat behind, with the Capponcinis and the Young Changs. Kate and Miranda measured each other with sideway glances, but Miranda didn't ask how James was faring, and Kate didn't volunteer any information.

Suddenly there was music; the beat of kettle drums

joined by the humming from the crowd. The anticipation heightened as Sorciero started a short incantation. Then he held out his arms and called for silence. He spoke to the people of Tequila, telling them of the abduction of young Charlie by their old enemy, the Curria clan.

'We thought that we had beaten those evil forces who would have taken over our beloved island some years ago, but the serpent reared its ugly head again, and tried to destroy us. They kidnapped our precious son Charlie, and tried to ransom him for Petrolia and all we possessed. With the help of God, our sacred spirits, and our loyal friends here in Tequila and neighbouring Mangorenia, we were able to crush our enemies. We freed Charlie and God punished the kidnappers. The ocean swallowed them up in a ball of fire. Long live our beloved Tequila and all who live on this blessed island.'

The crowd shouted with jubilation and the music started again. Sorciero twirled around slowly, moving his arms like the wings of a bird. The leopard coat fell off his massive shoulders, and the girls gaped at the sight of his bulging muscles. The Mère Magot rose from the Mandarin chair, and handed the crystal ball to Old Crone. She spread out her arms and ululated, invoking the spirits in her French Patois. Sorciero stepped forward and put his foot on the burning path, to the rhythmic clapping of his audience. When he was halfway he stopped; he looked to the starry heavens and continued to wave his arms like the wings of a bird. Cocorico screeched and left his mistress' shoulder. The bantam landed on Sorciero's head and flapped its wings in unison with Sorciero's arms. There was a great gasp from the audience as his feet started to leave the burning path, and he hovered above the ground, then slowly rose into the night sky. Maria Beatriz started up and grabbed his feet, to stop him. She implored him to return to earth, but he fluttered like a giant moth above the heads of the now wildly cheering crowd, lifting her towards the heavens. She let go of his feet and found that she was weightless. She held out her hand and he took it, and together they swayed above the crowd.

'An amazing act of levitation,' Jason whispered into Kate Harvey's ear.

'How ever did he do this?' Domitilla asked.

'It's like the Indian rope trick,' Mei-Ling murmured.

'It's mass hypnotism,' Young Chang said decidedly.

'It may be another property of the Mangorenian water they've been serving here tonight. Hallucinatory properties,' Maurizio added his opinion.

'Did any of you drink the water?' Miranda asked.

'I think we all had a sip,' Maurizio said. They all looked at each other in wonderment. Clearly it could be the water.

'Sorciero has attached himself to a balloon with fine plastic wire, and someone has released it.'

'You've got a wild imagination, Papa,' Miranda giggled.

'I think we have to accept the fact that our Sam has the gift,' Mei-Ling said. 'Besides, I think that the Mère Magot is a pretty powerful Shaman in her own right. Together they make great magic.'

The music swelled, got louder and more frenetic. The people started to dance around, waving their arms like the wings of a bird. Suddenly they were all hovering above the ground, screaming and yelling. They were doing summersaults in slow motion in the air, emulating astronauts in outer space. This phenomenon lasted for roughly five minutes and then it was return to earth and perfect quiet. The music had stopped, and only the sighs of the people as they landed on their feet disturbed the air.

'I felt like an astronaut,' Miranda said. 'Did you feel the same?'

'I never felt anything like that before. This great sensation of elation and weightlessness; did this really happen?' Maurizio questioned.

'It was orgasmic,' Jason Reed murmured.

'So it happened to you as well!' Maurizio exclaimed.

'And me,' Young Chang confessed.

The women looked at each other in embarrassment. Then they nodded and giggled like naughty children.

They all looked up and saw how Sorciero and Maria Beatriz landed and he continued walking on the burning path to the end. Cocorico flew back to his Mistress who stroked his ruffled feathers. Old Crone had closed her eyes and seemed to doze. Etienne and Marthe Bonheur had landed on the roof of the old cottage, and stood there hand in hand. A ladder was brought by the firemen and the guests were guided off the roof to safety. A great stillness had spread over the people, who were completely mystified by the experience of mass levitation and orgasm. It remained the strangest occurrence of any Happening in the history of Tequila, and speculation as to the cause was rife but no theory was ever rationally proven or established.

As for the cocktail of Mangorenian water and syrup, the effect was amazing on the people who had an opportunity to drink it. Their sexual prowess was massive and enduring that night. It was the talk of the paseo the next day, where the men stood around gossiping in low voices and sniggering like schoolboys discussing their first sexual experience. It remained a mystery to them why this should have happened, and they put it down to Sorciero's magical mystery levitation. They imagined, to their great sorrow, that such prowess wouldn't repeat itself until next year's Happening.

Chapter Eleven

Jason Reed and Kate Harvey returned to the consulate. They had been politely invited to join the others for dinner at the Club, but Kate had declined, wishing to return to James' bedside. All the staff had been given the night off on the occasion of the Happening, and only a marine was on guard duty. The table had been laid for three in the dining room, and a cold supper had been spread on the side-board. Kate Harvey knocked on the bedroom door and went in. James was sitting up in bed reading a paperback he had found in the bedside drawer. He looked up and smiled at his mother.

'Hi, Mom, did you have an interesting time?'

'It's hard to describe or even explain what took place at the Happening,' she answered.

'I've been told that all kind of strange things happen on that night.' He grinned, and put his book down. 'Did you bring this book here from home, or buy it at the airport?'

'No, Darling, someone else must have left it here. Is it any good? What's the title?'

'"Hot Water". It's very odd, because it's about life on an island rather like this one. You should read it, although it is a bit vulgar at times. It's made me laugh.'

'Do you feel well enough to join Jason and me for dinner?'

'Yes Ma'am, just give me a minute to slip on

something comfortable, like Mr Reed's dressing-gown, which he has kindly lent me.' He looked at his mother and thought how young and happy she looked. 'Did you by any chance see Miranda, or meet her?'

'Yes, to both these questions.'

'She didn't ask about me?'

'She didn't ask about you.'

'She sure knows how to keep a grudge, that girl.'

'You can hardly blame her, James. After all, you came here to harm her and her parents.'

'Does she know that Curria was never my father? Did you tell her?'

'The opportunity never arose. Besides I think that you should tell her.'

'Why should she believe me?'

'Because we will ask for a DNA test to prove it.'

'Don't you think that she's beautiful, Mom?'

'That she is, James, but then you're no dog yourself.' She patted his hand. 'Don't worry about it, Baby, everything will sort itself out. See you in the dining room.'

Jason stood at the side board and served Kate. She had never had a problem with weight and enjoyed her food. James pulled her chair back and she sat down. Jason poured the wine, but James refused. He never drank alcohol before a bullfight. He toyed with the food on his plate, eating sparingly.

'Mr.Reed, could you perhaps have my stuff brought here from the Club? I need my clothes for the bullfight tomorrow.'

'Please call me Jason, and yes, I'll send my driver for your luggage in the morning. How are you feeling?' Jason asked.

'Thank you for your concern, Jason. I'll be fine for the fight tomorrow. I've no headache, and my vision is normal.'

'Nevertheless, the Doctor will pop in tomorrow morning early to take a look at you.'

After dinner James went back to bed, and Kate Harvey tucked him in like she used to when he was a small child. He smiled contentedly and closed his eyes. She went back to the sitting-room and accepted the offer of a drop of cognac from Jason. They looked at each other speculatively, and when he reached for her, she let herself be held.

'I haven't slept with a man for a long time, Jason,' she murmured.

'Don't worry, Kate, it's like riding a bicycle; once you've learned how to do it, you never forget.'

'I trust it would be more satisfying than riding a bike,' she said, and laughed softly. 'But perhaps we should have another splash of the cool Magorenian water.'

'I reckon we really don't need it,' he said, and tried to kiss her fully on the lips.

'Don't take it amiss, Jason, but I need a little more time.'

Miranda sat with her parents and the Young Changs on the dining terrace of the Club. There too, a cold buffet had been set out for the guests. It was the one night of the year where guests had to help themselves, as most of the staff had been given the night off. Miranda drifted off into a dream world, in which she was being held in James' arms. Her anger had started to abate, and although she hadn't quite forgiven him, he didn't appear to be as big a monster as he had seemed to her a couple of days ago. He had, after all come clean and shown that he wasn't a complete villain. It amazed her that her parents had accepted his explanations, and forgiven him so promptly. It was always poor James this and poor James that, now that he was laid low with concussion.

'Hello, wake up, Miranda; are you going to eat something?' she heard her mother ask.

'No, Mama; I'm not very hungry.'

'Well come to the buffet anyway; you might just be tempted by something.'

Miranda obeyed reluctantly, and accompanied Domitilla to the buffet.

'You must stop worrying about James, Sweetheart. He's going to be fine.'

'I'm not the least worried about James,' Miranda tossed her head. 'Besides how do you know he's going to be fine?'

'I asked his mother. You could have done the same thing.'

'I'd rather tear my tongue out, than ask about him. Is he going to fight tomorrow?'

'Yes, he certainly wants to.'

'But he's in no condition to face the bulls. They mustn't let him.'

'It's his decision, Miranda. He must know best how he feels.' Miranda's eyes filled with tears, and Domitilla hugged her.

'Dry your tears, Randy, and put a smile on your face. It's time you made it up with James.'

'How can I, Mama, when he's going to be gored to death by a bull tomorrow?' she wailed and hid her face on her mother's shoulder.

Sorciero and Maria Beatriz were entertaining Etienne and Marthe Bonheur at a late cold collation in the Presidential villa. They too, were helping themselves, as the staff had been given the evening off to go to the Happening. The prevailing mood was contemplative. The phenomenon which had occurred at the Happening had had a strange effect on the party. They were able to talk of nothing else; even the daring rescue of Charlie was put in the shade by the mass levitation which had taken place that night. They questioned Sorciero again and again, but he could find no rational explanation. Indeed he hardly remembered it. All he knew was that he had found himself almost above the great treetops, when he came out of his trance, with no idea how he had got there nor what was keeping him floating. He was holding Maria Beatriz' hand, and feeling as horny as hell. He could see all the people turning summersaults, and space-walking below him. He could have sworn that he'd seen the Mère Magot cruising around on a broomstick, grinning from ear to ear with her best dentures. It was an awesome sight which had made him shiver. He had closed his eyes and prayed to the spirits of his ancestors, imploring them to bring him and Maria Beatriz down to earth gently. Naturally he didn't tell his guests about the vision he had had about their Shaman, and her means of locomotion. To Sorciero's relief, she had elected to take her meal in the nursery with Charlie and the nanny. Cocorico and Charlie had taken to each other and were inseparable. The door to the dining room burst open and Charlie skipped in with the bantam sitting on his shoulder.

'Papa, Papa, Mrs. Maggot said she would teach me to ride in the sky on a broomstick. That sounds much more fun than riding a bicycle,' Charlie cried excitedly. 'Where can I find the broom, Mama? We could start my first lesson right away, couldn't we?'

Chapter Twelve

Palm Sunday dawned serene on the island of Tequila. There had been a few showers before dawn, but now the rising sun had torn apart the thin layer of cloud, helped by the gentle breeze. The island stirred late; the activities of the night before had exhausted the people, and everyone had slept in. Shops would remain closed today, and there would be a Te-Deum at the cathedral at noon. It had become traditional that the bullfighters would come to the service and be blessed by the bishop, before entering the arena. The congregation started to fill up the Cathedral, which was awash with fresh bouquets of Easter Lilies. The scent was almost overpowering. The women wore their beribboned straw hats and the customary white, and the children looked neat and laundered. The diplomatic corps attended in full force as did the members of the Government. Sorciero sat in the front row with Charlie perched on his knee.

The organ and the choir had started the first hymn, and were presently joined by the congregation. The choir master was conducting, and the music soared into the high dome, reverberating through the antique timbers. The bishop said Mass, and then held his sermon. Now came the moment they had all been waiting for. The bullfighters came through the great doors, led by Maria Beatriz. They trod down the aisle with measured steps, resplendent in their embroidered finery. Each in turn took Holy Communion and was blessed by the bishop.

'Where are the bulls and the horses?' Miranda whispered to her father, her eyes shut tight.

'This isn't the Palio in Siena, Randy,' he whispered back.

'Doesn't James look splendid in his costume?' Domitilla gave her daughter a sideways glance, and nudged her gently.

'Stop nudging me, Mama. You're hurting my ribs,' Miranda said, and closed her eyes even more tightly.

'He's staring at you with such intensity, it's a wonder you can't feel it. Don't you feel the least bit sorry for him?'

'I'm not the least bit sorry, so there,' Miranda retorted.

'That's what you used to say when you were little, Darling, in exactly the same voice, when you were asked to apologize.'

'But I'm all grown up now, Mama, and still I can't forgive him.'

'You can open your eyes now, they've all retreated to the back pews. It was quite a spectacle. Pity you missed it.'

'Can I come to the bullfight with you and Papa?'

'I thought you didn't want to see James?'

'I don't, close up, but in the arena, that's different.'

After the Mass at the Cathedral, the Tequilans began their trek to the Arena. Those who had tickets were shown to their seats, while others waited in line for the last available seats. As usual, enterprising youths had climbed electricity poles and were perched perilously on roof tops and balconies of the neighbouring houses, in order to get a glimpse of the proceedings. Three fire engines were standing by, and the ladders had been extended so that more spectators could cling to them. On the way, people had stopped at the now busy food stands. They went away with sandwiches stuffed with grilled sausages, shrimp, or grouper. The local sausage factory had contributed the sausages. The fishermen had had a bumper catch, and the whole catch had been bought and donated by the Government for a free meal. No alcohol could be bought on that day, but the lemonade and other soft drinks stands were very busy.

The Mère Magot had been given a seat in the Presidential box. This time she wore a nun's habit which she had inherited from a nun who had died in the convent of St. Helena in one of the French Antilles. It was there that the Mère Magot had learned to read and write. The habit had a resplendent starched white collar, and a trapeze shaped white coif. Cocorico was also present, and had perched itself on Charlie's little shoulder. The President of Mangorenia and his wife were elegant in pristine white linen. Old Crone Chang was in her ubiquitous black trouser outfit, and Mei-Ling wore a white brocaded chungsam and her pearls. Her shiny black hair was twisted into a chignon held in place by pearl-studded tortoise shell combs. Both she and her mother-in-law, held small parasols to shield them from the midday sun. The Capponcini family sat together, dressed in the traditional white. Domitilla and Miranda wore white lace mantillas, as did most of the women in the arena. Kate Harvey was flanked by Jason Reed, who surreptitiously tried to hold her hand under the cover of her shawl. She was wearing wrap-over sunglasses which hid the fear in her eyes. She had never attended the bullfights which were fought by James. This time it was unavoidable, and she felt almost sick with anxiety at the impending spectacle.

A fanfare of trumpets sounded and the spectators knew that the first bull was about to be released into the arena. The horses, wrapped in their protective covering, were ridden in by the picadores and dispersed in the ring. The gate to the bull pen was flung open and the bull stood framed in the gateway, slowly pawing the soft earth. His massive head turned from side to side, as the red capes were twirling in his field of vision. Suddenly he lunged forward and made for the nearest man holding a cape. The fellow quickly withdrew behind the wooden panel, while the bull butted it furiously. He turned back into the ring, and now the riders came forward to taunt him. He charged the poor horses repeatedly, always being hit by one of the lances of the riders as he did so, to the cries of 'Ole' from the crowd. Blood was beginning to leak from the wounds in his body, and he shook his head as if to clear his vision. The horses departed, rather the worse for wear, and it was time for the real business to start. A young Torero from Spain was about to show his prowess, and the crowd cheered and whistled as he executed his first Veronica. It was also to be his last. He slipped on the bull's excrements, and was lucky to be able to make his escape, covered in shit. Now there were boos and whistles from the crowd, and the bull lifted his head and pawed the earth menacingly. Then he lifted his tail and let out another river of shit, just for good measure. Titters of laughter were now heard and the crowd started to slow clap. The organizers decided to withdraw the animal which obviously had some form of stomach disorder. They had a cow on heat in reserve for just such an emergency, and she was chased into the arena. She sidled up to the bull and then trotted back to the entrance of her pen, swishing her tail. This galvanized the bull into action, and he raced after her and mounted her with apparent gusto. The crowd was delirious by now, and the screaming and yelling was deafening. Etienne Bonheur wiped the tears from his eyes, and Marthe Bonheur managed to look slightly shocked at the spectacle. Sorciero was standing at the rail urging the bull on.

'Give it her good, Mun' he yelled, jumping up and down. 'Show her you've got balls, Mun.'

'They must have watered the bulls with the mineral water,' Jason murmured into Kate's ear and tried to stroke her thigh.

Stable-hands were now busy cleaning up after the bull, and strewing fresh sand. When all the excrements had been removed, the fanfare announced the second round. The crowd settled down and the ritual started again. This time there were no mishaps and the young Tequilan Torero acquitted himself honourably and was awarded the tail of the animal. The next man came from Venezuela, and his bull was definitely not to be taken lightly. The Torero narrowly escaped being gored, and after numerous tries, finally managed to plunge the sword into the bull's neck and bring the contest to a gory, if not glorious, end. The stable-hands ran into the arena, again raked the sand and picked up the debris which had been thrown into the ring. Now it was time for one of the highlights of the afternoon. The people sat forward attentively and a frisson went though the crowd as Caramba de la Cruz walked into the arena and doffed his hat to the presidential box. He was incredibly handsome and elegant in his costume, which fitted him like a second skin. His fair hair, twisted back in a queue, shone golden in the sun. He retired behind one of the wooden boards and waited for the bull to be let loose. This time the bull came charging out, snorting loudly.

He chased the horses mercilessly around the arena, avoiding most of the lances. The horses withdrew and it was time for James Harvey to show his mettle. He walked towards the bull, tossing his cape, daring the animal to come and attack. The bull didn't hesitate, and attacked the twirling cape. The first Veronica had been completed faultlessly and with elegance, and the crowd cried 'Ole' with every one that followed. Then the banderillos were placed into the bull's neck, and the beast seemed to tire at last and start to buckle at the knees. It was time for the 'coup de grace', and James approached the bull, the sword hidden in the 'cape'. They lunged at each other, and the sword was implanted successfully in the neck. The crowd cheered wildly, and threw everything they could lay hands on into the ring. James turned his back on the enemy and walked away. He looked up at the Presidential box, and saw Miranda waving wildly, blowing kisses in his direction. He was so taken aback by this unexpected manifestation, that he didn't hear the last attack of the dying bull as he plunged towards James, butted him in the back, and gored him. Another Torero sprang into the arena and approached the bull, challenging it to fight. It was Sorciero who recognized Maria Beatriz and then everyone was standing calling her name. She managed to attract the attention of the bull. She sank her sword into his huge neck, and stood aside to watch him die. Now the arena was alive with first aid and ambulance men. They loaded James onto a stretcher and ran with him to the exit. Miranda jumped into the ring and ran after them. She managed to hop into the ambulance as it was leaving for the hospital. Kate Harvey held on to Jason, looking ashen, and Marthe Bonheur took a bottle of mineral water from her ample shoulder bag, and splashed into Kate's face.

In the arena, Maria Beatriz was hoisted onto the shoulders of the fans, and carried around triumphantly. She cut a dashing figure in her turquoise blue and gold costume, and waved to the Presidential box, where Sorciero was lifting Charlie onto his shoulders. As she approached the box, the boy was handed across to his mother, and continued the victory ride with her. It was another great day for the President of Tequila, as the people cheered and sang the national anthem.

James lay on his stomach on the hospital bed and winced as the wound was disinfected and the stitches were put in by the surgeon. Miranda waited outside, tears streaming down her face. Domitilla tried to comfort her, and Maurizio was urging her to take a sip of brandy from his hipflask.

'He's not going to die, is he, Mamma?'

'Of course not, Darling; it might be just a superficial wound.'

'Why don't they tell us anything?'

'We're not next of kin.'

'Of course we're next of kin, I am going to marry him, and you'll be his mother-in-law.'

'This is a new turn of events; thank you for letting me know!' Domitilla said sternly. She raised her head and saw Kate Harvey running into the corridor, accompanied by Jason Reed.

'Where is he?' Kate cried.

'In the operating theatre.'

'Oh God, let him be alright,' she looked at Miranda. 'You were in the ambulance with him, they tell me. Was he conscious?'

'Yes, but in some pain; he seemed to be losing some blood,' Miranda answered and wiped her eyes.

'He loves you very much,' Kate Harvey said and patted Miranda's cheek. They all started forward, when the door of operating theatre opened and the surgeon came out.

'Quite a gathering,' he commented.

'How is my son?' Kate Harvey cried.

'He was lucky. The bull caught him in the buttock and not the small of his back. We've cleaned him, stitched him up and given him an antibiotic and a tetanus jab. He'll find sitting down a wee bit uncomfortable for a few days, but once the stitches are out, he'll be as good as new.'

'Thank God,' Kate Harvey exclaimed. 'Can we see him now?'

'You can see him and take him home as well in a couple of hours. We need to give him a transfusion, just to be on the safe side, because he has lost some blood. After that, he's all yours.' He looked at the mother and asked: 'Are you the same blood group as he is?' Kate shook her head regretfully.

'I'm just an ordinary O something but he's AB negative, quite rare I think,'

'I'm an AB negative aren't I, Mama? I can give him blood,' Miranda said eagerly.

'That's fine, we'll just test you, and then you can go in and do your worst.'

It took a little time before Miranda's blood was pronounced the right one, and she was taken in, and lay down on a bed beside James. He turned his head and grinned at her. 'So you're the donor?'

'Stop grinning like that, James, this is serious business.'

'You'll never escape me now, Randy. This has blooded us together for life. You realize that don't you?'

She made a foul grimace at him. 'Vampire,' she hissed and took hold of his hand.

James fell asleep with a smile on his lips, and

Miranda watched his even breathing. The blood transfusion had been performed without any trouble, and she was sipping a glass of fresh orange juice to restore her. Kate Harvey had come in and sat on the edge of Miranda's bed. She had recovered from the shock of seeing her only child gored by a bull. Miranda studied her, and found that Kate was still a most attractive woman. She hoped that they would be friends, because nothing could come between her and James now, not even his mother, or hers.

'I see that you have forgiven James, my dear.'

'I don't know whether I have forgiven him, Mrs. Harvey, but I do love him so much that it doesn't matter.'

'Please don't stand on ceremony, just call me Kate,'

'Alright, Kate it is. You know we're going to get married,' Miranda said simply.

'I expect you will. You're both very young, but very determined; I can see that.' Kate responded kindly.

'We'll be separated for some time, one way or another. I want to graduate from university before we have any children, and James will be busy with the ranch and his bullfighting.'

'I'm glad that you are approaching this in a sensible way, Miranda,' James said sleepily and opened his eyes. 'I, however, have other plans. I don't want to be separated from you, I don't like long engagements and I don't want to sleep with the girl I marry before the wedding. So you see, my precious, there will have to be a compromise.' He held out his hand to her. 'So what are we going to do?'

'How about getting to know each other better,' Kate Harvey ventured. 'There are some things that James will have to tell you about, Miranda,'

'Well it can't be worse than what I already know.'

'You're right, my Love. Isn't she brilliant, Mother?

Not only beautiful but smart. How long do I have to stay here, Mother?'

'The surgeon said that you could leave as soon as you woke up.'

'Where shall I go?'

'I think you had better come back to the consulate. The club is a bit far to ride to on your knees, because you can't sit down. The dressing will have to be changed tomorrow and the day after, here, at the hospital.'

'Alright, Mother; I'm sorry to give you all that trouble.'

'Don't worry about that. The main thing is that you're not seriously hurt. I'll tell the nurse that we want to leave. Jason will drive me to the consulate and I'll pick up something for you to wear. Miranda will keep you company until I get back. How does that grab you?'

'It grabs me fine!'

James looked at Miranda adoringly. He had managed to turn over on his side and was facing her. He held her hand and kissed her palm, then all the fingers. She had nice hands, strong but slim and long fingered with well shaped nails.

'Stop that, James,' she murmured. 'It's making me feel,'

'It's making me feel too.'

'You've got lovely hands,' he said and continued kissing and licking her palm. She brought his hands to her mouth and started sucking his thumb, curling her tongue around it.

'Come and lie down next to me,' he said pleadingly.

'I thought you said no sex before marriage.'

'This isn't sex, I just want to feel your warmth, next to me.' She smiled and snuggled up to him. He kissed the tip of her nose before travelling down to her mouth. She opened her lips to him and he explored her mouth.

'God,' he moaned, 'You taste so sweet, Randy.' She hid her face in his shoulder and pressed closer to him.

'I bought your engagement ring a few days ago, at Young Chang's,' he whispered into her ear.

'What is it?'

'An emerald, plain and simple.'

'Mmm... my favourite. How did you know that?'

'I know all about you... I know you inside out. I know what your breasts look like, your brown tipped nipples, the silky curls between your legs.' He stroked her thigh. 'Jesus, this is like being in purgatory, and being denied entry to heaven. How long will you make me wait?'

'Are you asking Jesus, or me,' she queried. 'I'd better go back to sit on the chair. Your mother might come at any moment with your clothes.'

He sighed and released her reluctantly. She got up, smoothed her short skirt and sat down in chair again, her hand in his.

'Your mother said you had certain things to tell me,' she said. James closed his eyes again and sighed.

'Is it that bad?'

'Not really, my love, it's actually good news. It's just a bit complicated. You see, Romeo Curria wasn't really my father.'

'And your mother isn't really your mother; you certainly don't look like her. Let me guess... You were a foundling, you were adopted.'

James laughed, and then pulled a face. 'Nothing as romantic as that. I mustn't laugh, it hurts; but no, I wasn't adopted.'

'Poor baby,' Miranda commiserated. 'So who is your father then?'

'My mother's first husband. They were high school kids, very young rather like us. They eloped to Las Vegas and got married there. Then his parents found them, and demanded that the marriage be annulled, because they were not of age. My father's family were stinking rich, and my mother just an orphan from the wrong side of the tracks.'

'How awful! What happened then?'

'My father was too young and too weak to oppose his parents, and my mother decided to run away and got a job on a cruise boat, as a stewardess. After a few weeks she found that she was pregnant. On board ship she met this good-looking, obviously wealthy Italo-American, who fell in love with her. She needed a father for her unborn child, and when he proposed, she agreed to marry him and live with him on his ranch in Argentina.'

'So she married Romeo Curria?'

'Yes, but she found out that he was already married, and that he was a mobster. He couldn't have children with his lawful wife, so he was pleased to have a son. Me.'

'And what happened then?' she asked.

'He terrorized my mother and made her stay at the ranch. She agreed because she thought it would be the best thing for me. A quiet life in healthy surroundings, a good home, an education and no money worries if she toed the line. That's all really.'

'That's not all, that's really something, James. Thank God that it's worked out that way. There's not need for you or your mother to be afraid anymore, now that the threat of Romeo Curria is no longer.' She looked at him lovingly.

'My only worry is that there may be another Curria lurking round the corner,' he said.

'Even if there is, he certainly couldn't claim to be your father,' she giggled. 'Will your mother now tell your real father about you? It would be wonderful if the two of you could meet. By the way, what's your real father called? Did she tell you?'

'His name is Bruce Winthrop the third, or fourth. The family goes back a long way.

Miranda frowned and looked thoughtful.

'What's the matter, my love?'

'The name sounds familiar; I'm sure I've come across it quite recently.'

'You might have seen it in the press. The Winthrops own a huge chemical company, my mother says.'

'I don't usually read the financial press.' She got up and walked up and down the room. Suddenly she clapped her hands and cried: 'I've got it. You won't believe this, Darling, but someone arrived at the club today called Winthrop. I saw the card in the office, but I didn't see the person.'

'It's a coincidence, it just couldn't happen,' James said breathlessly.

'But suppose it's him, Bruce Winthrop the fourth, and all that. How cool would that be?'

'Shivering cool. I've got goose bumps.'

'I'm going to phone the club, right now.' Miranda dug into her shoulder bag and found her mobile. She got through to the club and spoke to the reception, asking them to check on the name. She held on for a few seconds and then nodded and nodded again, saying 'aha' several times, before ringing off.

'Well for God's sake, don't keep me in suspense!' James cried out. She smirked and wrinkled her nose at him.

'Do you really want to know?' she teased him.

'You're really brutal, my love; here I am, injured and suffering, helpless. So stop torturing me.'

'His name is, let me see... I think I've forgotten it already.'

'Stop, or I shall go mad.'

'His name is Bruce Winthrop from Boston. Company Director of Winthrop Chemicals.'

'This is just too much,' James whispered. 'Too much for me to take in, on one day.'

'What about your mother, shall we tell her?'

The door had opened, and Kate Harvey stood there with Jason Reed.

'What are you going or not going to tell me?' she asked James and Miranda looked at each other in silent communication and remained silent.

'Is anything wrong, James?' Kate asked anxiously.

'Not exactly, Mother, but you have to brace yourself for some news. We know for a fact that a Bruce Winthrop the fourth from Boston has checked into the Club Royale today.'

'Oh damn,' Jason Reed exclaimed and supported Kate Hudson who had fainted dead away. 'Why do these things always happen to me?' he complained. 'Just as we were getting to know each other. Had we better get a doctor?' He picked up Kate and carried her to a chair.

'Jason, you really shouldn't try and seduce my mother, now that it appears that my father has so miraculously turned up.'

'I'm not giving her up so easily, James,' Jason said decidedly. 'Your father is probably there with his wife and six children, and what good is that to your mother?' Jason patted Kate's cheek, and Miranda stood by with a glass of water.

'Wake up Kate, that's a good girl. Open your eyes now.' Kate Harvey moaned and put her hands to her head. 'What happened?' she murmured.

'You had a short fainting spell. Here, have some water, Kate; it'll do you good.' She took the water gratefully and drank. She raised her eyes to James and said: 'Is it true James, or did I dream it? Did you say that your father is on the island?'

'It would seem so, Mother. He's registered as Bruce Winthrop, CEO of Winthrop Chemicals at the Club Royale.'

'How stupid of me to faint, just because I heard his name,' she whispered. After all these years, her heart still fluttered when she thought of him. Her knees felt weak at the mere thought of him being but a few miles away.

'I want to go and see him, Mother. Not tomorrow but now, even if I have to be on my knees for the drive. Will you take me?'

'Yes my darling boy; I'm just as anxious as you to see him.'

It was already dark as they left the hospital and drove back to the club. Jason had offered to drive them in his car, as Miranda's Landrover was declared to be too uncomfortable. James was lying on his stomach on the back seat, and Miranda followed in the Landrover. The town was brightly lit and the Tequilans were out in full force enjoying the balmy evening. Those who had been invited to the Ball at Government House were starting to get into their finery. It was the social event of the year, and after the dinner dance a splendid firework display over the harbour was put on for every one to enjoy.

Jason pulled into the Club driveway and stopped at the reception. James managed to crawl out without injuring himself. He wore a T-shirt and a sarong draped around his hips. His flaxen hair was still tied back in a queue. He was immediately surrounded by admirers as he entered the hall. The dire end of the corrida hadn't damaged his image in the least. It was considered the height of romanticism, that James had ignored the threat posed by the bull, in order to salute his beloved. A small group of English children who had arrived with their parents for the Easter holidays stood by and gawped.

'Look Mummy, it's Becks,' a small voice piped up, 'but she doesn't look like Posh at all, does she?'

'It's not Becks, darling. This young man is a bull fighter called Caramba de la Cruz.'

'But he's wearing a skirt and has his hair done like Becks, Mummy,' the child protested. 'I'm going to ask for his autograph anyway, just in case.' Miranda went to the desk and asked whether Mr Winthrop was in his cottage. Jolie told her that the guest was having a drink at the outside bar. James smilingly signed autographs, and slowly walked on to the terrace, accompanied by Miranda and Kate. He scanned the guests sitting at the bar, and squeezed his mother's arm as his eyes rested on the man who was sitting on a bar stool, and was staring out to sea. His hair was as fair as James', and it glistened white gold under the muted lighting.

'Is it him, Mother?' James asked quietly.

'I do believe it is. Of course I can't be quite sure; don't forget haven't seen him for over twenty years.'

'Shall we just stroll by him, nonchalantly, as it were?'

'Must we?'

'We must!'

'Supposing he isn't alone...'

'Oh but he is,' Miranda murmured behind them. 'I checked in reception.'

James started steering Kate to the bar. 'Be brave, Mother, and look him in the eye,' James encouraged her. Slowly they began crossing the man's field of vision, and James stopped and said; 'Sorry if we're spoiling your view, Sir.' The man abandoned the horizon and the sea, and looked at James. The resemblance between the men was uncanny. The same cut of the jaw, the same bright eyes and soft, gentle mouth.

'The view is for every one, and I was just going anyway.' He stared hard at James. 'Have we met before? Your face seems so familiar.'

'That's because we look very much alike, Sir,' James smiled at him and turned to Kate. 'Don't you think so, Mother?'

The man got up from the barstool and turned to look at Kate. They stood staring at each other wordlessly.

'Forgive me for being so rude, I think I must be dreaming.' The man turned away and leant on the bar.

'You're not dreaming, Bruce. We're wide awake in Tequila,' Kate said quietly.

'I've been looking for you for years, and now when I had given up all hope, there you are.' He shook his head. 'It's a dream it; must be. And this young man who looks so like me must also be a dream.' He turned again and touched her cheek, and then James' in wonderment. 'Maybe you are real,' he said, tears starting to gather in his eyes, and then they were all crying.

'Let's get away from here, I don't want to share you with all these people,' he said and took them both into his arms. Miranda retreated to the archway of the reception where Jason and Maurizio Capponcini had been watching the scene.

'Just my luck,' Jason said glumly. 'She was really getting interested in me, and then this. One chance in a billion that this guy would turn up, and he does...'

'Let's have a drink, Jason, to cheer you up.'

'As long as it's not mangorenian mineral water. I feel somehow that it's not going to be the right night for it, after all.'

Chapter Thirteen

The old adage that the path of true love never did run smooth certainly appeared to have some basis, if one considered the Winthrop-Harvey affair. It had been a peculiarly bumpy road; a real roller coaster of a path. The three personages involved were still holding hands, strolling down the path, with a view to ending the rough ride, and to glide into the safe harbour of happiness and tranquillity. They reached Bruce Winthrop's cottage and settled down on the deck, except for James who leant against the pillar. The hurricane lamps had been lit and cast a soft glow on them. Bruce Winthrop continued holding Kate's hand, and looking from Kate to James, he shook his head in amazement.

'If I am asleep now, and if I should wake up and find it was all a dream, I shall just wade into the sea and disappear,' Bruce Winthrop said and touched her cheek hesitantly. 'You are so beautiful Kate, more beautiful even than when I first saw you.' He turned his gaze onto James and his eyes were bright with excitement. 'He's mine, isn't he?' It was more of a statement than a question.

'Well, technically, he's ours,' she teased him gently.

'How could you, Kate? Just disappear and never tell me about him?'

'What else could I have done? The annulment gave me no choice.'

'I know that it's all my fault, my love, I should have stood up to my folks.'

He jumped up and started pacing the deck. 'I don't know what I'm saying; calling you my love, assuming that all this time you've just been waiting for me to turn up.' He glanced at her, sitting calmly in her chair. 'You're here with your husband I assume, and are happily married, have other children.'

'I assume that you're here with your wife, and other children?' she countered.

'I'm divorced; have been for a long time. No children... except him. Where have you been hiding him all these years?'

'In Argentina, on a cattle ranch. We also breed bulls, and James is quite a famous Torero. His name is Caramba de la Cruz.'

'So you're the darling of Tequila who got injured this afternoon. The club was humming with rumours that you had been fatally gored. Thank the Lord, that isn't true.' Bruce Winthrop looked searchingly at Kate again. 'You didn't answer my question, about a husband and a horde of kids.'

'It's a long story, and I think that we need time to tell each other everything,' she answered.

'Lord, woman, don't torture me,' he cried. 'It's a simple enough question to answer.'

James smiled at them, and put his oar in. 'I'm going to my cottage. I feel tired and must lie down on my bed. Randy will bring me some supper. You can share my cottage, Mother. No need to go back with Jason to the consulate.'

'Who is Jason? What consulate?' Bruce asked.

'Jason Reed is the American Consul, and I am a house guest at the consulate.'

'James is perfectly right, Kate. No need to go back with this Jason; you can share my cottage. In fact, I am asking you to have dinner with me, here at the Club. I have a table reserved on the terrace.'

'But I have date with Jason to go to the Government Ball; I don't know whether I can break it just like that,' Kate objected.

'God, I do take everything for granted, don't I? Forgive me Kate, but it's so strange and wonderful to have found you, that I cannot let you go again.'

'It's been a long time Bruce, and we both need time to consider everything. I will have dinner with you, but spend the night in James' cottage.'

'That's right, Mother. Play hard to get. It always works,' James advised his mother. 'Listen to the voice of experience,' was his parting shot as he left his parents and slowly made his way to his cottage. He was beginning to feel the stitches, as the Novocain lost its painkilling effect. He was glad to reach his cottage and went in. Miranda was sitting in the armchair, with a tempting dinner laid out on the low table. The champagne had been poured and the bottle was cooling in the ice bucket. She rose to meet him and he folded her in his arms.

'That feels so good, just like coming home after a long and arduous journey,' he sighed. 'I'm tired and uncomfortable, and only a few days ago I would have wanted to be with my mother, and have her cosset me. Now I want you to cosset me.'

'Yeah? How does one do that curious thing?'

'I'm going to stretch out on the couch, sideways on, and you are going to feed me and water me with a little Champagne. Then you're going to give me the antibiotic and hold my hand until I fall asleep.' He grinned. 'That's cosseting.'

'A few days ago I would have told you where to get off, but tonight, exceptionally, I will do your bidding.' She grinned back. 'It's amazing how alike

Latin men are,' she continued and led him to the couch. He knelt down in front of the low table, and picked up a goujon of sole, dipped it into the tartare sauce and held it out to her, before popping the next one in his mouth. Miranda looked disappointed and frowned.

'Here I thought you were getting on your knees to propose to me, and all you wanted was to stuff your face,' she pouted.

'Oh my love, how insensitive of me. Why don't you look in the drawer of my bedside table and bring me the box you find there.'

She ran to the bedside table and opened the drawer. She picked up the night blue velvet box and brought it to him. He pressed the box and the lid sprang open. She gasped at the size and quality of the emerald which was cushioned on the velvet. It was mounted on a plain platinum band. He took the ring and she held out her hand. He slipped it on her finger. It was a perfect fit.

'Will you marry me my love, and live with me until death do us part?'

'I'm sure I don't know, kind Sir; you will have to get my family's blessing before I commit myself.'

'Oh cruel damsel, are you intent on breaking my heart?'

'Yes, every day of our life, but I promise to put it together again every night.' She knelt beside him and they rubbed noses, Eskimo style.

'Te quiero, amor.'

'Ti amo, my love.'

The Ball at Government House was about to start. Maria Beatriz had been checking the flower arrangements, the buffet and the bar. The waiters were at the ready with their trays, and the orchestra had started to play some Strauss waltzes. The ball always started formally with Maria Beatriz and Sorciero greeting the guests in the great hall. The guests were then requested to go through to the vast ball-room, and were offered drinks. Government House had been the old President's last great extravagant creation. Marble staircases, Onyx inlays, and precious wood had been used for the reception rooms. The kitchens and pantries were equipped with the latest electronic ovens, microwaves, dish-washers and cold rooms. A double sweeping staircase led to the upper floors where the guest suites were located. There were four guest apartments, decorated by the best American decorator, which were put at the disposal of foreign dignitaries who came on official visits to the Island. Tequila was a popular place for state visits. The guests were always received in style and they made the most of the weather and the gorgeous beaches which hemmed the island. This year, the Prince and Princess of Joli-Bonbons, and the prime minister of Mesopatamia and his lady, had been invited to stay at Government House.

When everything had been inspected and was to Maria Beatriz's satisfaction, the great doors were thrown open by two footmen and the first guests entered. Maria Beatriz was resplendent in scarlet palazzo pyjamas, a scarlet lame top, which bared some of her generous bosom, and a great scarlet lame bandanna tied around her head. The two ends of the bandanna flowed over her shoulders down to her waist. She had on a filigree gold collar which she had inherited from her mother, with matching ear-rings, which hung down to her shoulders. Sorciero had been poured into a dinner jacket, and stood by her, flashing his brilliant smile at each new guest. Young Charlie was perched on the banister at the top of the great staircase, and watched excitedly as the guests began to fill the hall. Behind him stood the Chinese nanny and the Mère Magot. The latter was resplendent in an exotic Chinese Mandarin full length coat, which she had borrowed from Old Crone Chang. These two had immediately formed a bond and discussed various medicinal herbs which they grew in their back yards. Mère Magot had arrived early with Charlie and the nanny. The boy had been given permission to see the arrivals and he was trembling with excitement.

'Look,' he pulled at Mère Magot's sleeve. 'Doesn't Mamma look beautiful? Just like a Princess in a fairy-tale.'

'Yes Young Charlie, she looks wonderful; tres belle. Ton Papa aussi. He looks very elegant.'

'Papa looks weird in this suit. Not like he usually does,' the child commented. 'Can I slide down the banister to give Mamma a kiss?'

'No, child, she's too busy now receiving the guests. Besides, you're not supposed to go down. That was the bargain.'

'Then can I show them all my new trick? The one you taught me, Mère Magot?' he pleaded.

The Mère Magot shook her head decisively. 'You haven't practiced enough, Young Charlie. You're not ready yet.'

'Oh yes I am,' Charlie said and ran into the corner where he had rested his wooden, painted hobby horse. 'Shall I show you?' Before they could stop him, the child had taken the hobby horse between his legs and taken off from the top of the stairs.

'Look at me, everybody,' he cried and circled the hall, riding the hobby horse high above the crowd. Maria Beatriz clutched her bosom and looked ready to faint. Sorciero was holding her up, his own knees like jelly. The guests were screaming in alarm, fearing a fatal accident. The band stopped playing and all the staff gathered in the hall and gazed towards the ceiling. Everyone was mesmerized by this extraordinary display. There was no end to the surprises this Semana Santa had in store for the visitors from abroad. Young Charlie was waving to his captive audience as he flew around, giggling with excitement. Suddenly Cocorico flapped down from the top of the stairs and landed on Charlie's shoulder. It seemed to be a signal, because Young Charlie ended his exploit and flew back from where he had started. There was a stunned silence in the Great Hall, and Charlie saw the sea of wide eyed faces staring up at him.

'Charlie, how could you,' Maria Beatriz wailed.

'It was easy, Mamma. Mère Magot said I had the gift. She said something in French too. "Tel pere, tel fils." It means "like father like son."'

'One thing is sure, Baby,' Sorciero cried. 'Blood is thicker than water!'

The next morning dawned to intermittent showers. It took the sun a while to disperse the clouds and win the contest. The countryside looked freshly washed and cool, and the smell of sweet Jasmine hung in the air. The sea was calm and the early birds on the beaches harvested the shells which had been washed ashore during the night. The harbour was busy. It was the annual 'Round the Island' Regatta Day and participants were getting their vessels ready. There were some hung-over, bleary-eyed faces around; the aftermath of the Government Ball and other celebrations on the island. Most people had watched the superb fireworks display at midnight, which had lit up the whole of the harbour again and again. The harbour had now been cleared of the fireworks pontoons, and was made ready for the huge cruise ship which was due at noon. It was the last day of celebrations before the advent of Baster.

Maria Beatriz and Sorciero were finishing their late breakfast in the company of their guests from Mangorenia. Charlie was munching his 'pain au chocolat', feeding Cocorico little pieces now and then. The little rooster pulled Charlie's hair gently to remind him. The incident of the hobby horse ride was still causing ripples of anxiety, and Maria Beatriz was not quite over the shock of seeing Charlie floating at ceiling height. La Mère Magot assured her that the child was in no danger, but that he would have to understand that he was not to squander his gift lightly. Such rides were to be used judiciously. She herself, Mère Magot, would speak to Charlie about it, and he would heed her lesson.

'Can't I even practice a little,' Charlie begged.

'No, Young Charlie, this exercise don't need practice. And listen good, Petit Charlie, if you don't do as I say, I will take the magic away, and you'll never ride again!'

'But Harry Potter does it all the time,' Charlie objected.

La Mère Magot opened her eyes wide. 'Je m'enfiche de ce Harry Potter. Who is he anyway?'

'He's in a book, Mère Magot, where they all make magic.'

La Mère Magot sniffed derisively and got up from the table. 'He's not real, then, is he, Petit Charlie, but you are. So take care and do as I ask you.' She stared hard at Charlie who now cowered in his chair. 'Viens, Cocorico, must go and meditate.' The cockerel left Charlie and landed on Mère Magot's shoulder. She strode to the door, but turned round before leaving the dining-room. She raised her bony finger and shook it at Charlie. 'Promise?'

'Promise,' the child answered solemnly.

The meeting had been set for 3 p.m. at the Presidential

Villa. Old Crone Chang, accompanied by her daughter-in-law Mei-Ling, arrived in Old Crone's rickshaw. The same lad had been driving her for eight years. He had been barely a teenager when she had engaged him. Now he was a grown man, and had a small fleet of rickshaws; it was a thriving little business, which Old Crone had helped him start. He, himself, now worked exclusively for Old Crone. He was her protector and she was his business adviser. Sometimes she was called out at night to administer some potion to a patient. He was always ready to take her where she wanted to go. His loyalty was exemplary, and he was a fine source of the rumours and cabals in the market place. No one had as good an intelligence service, as it were, than Old Crone.

'What's new in the town, Manuelito?' she asked as they passed through the gate of the Presidential Park.

'Not much, Old Crone. Everyone is still gossiping about the Happening, and Master Charlie's new trick. The one he performed at the Ball last night. Did you see it, Old Crone?'

'I did, Manuelito, it was sensational. That child is going to be a great witch-doctor.'

'He really did it without strings attached?'

'Not a single one, Manuelito.'

'I've heard that there must have been some drug used at the Happening,'

'Not exactly, Manuelito,' Old Crone smiled. 'You'll be the first to know about how it happened, if we ever find out. We'll get out here, thank you.'

Manuel, the rickshaw driver, stopped and helped the ladies out of the vehicle and handed Old Crone her basket. The guard at the doors of the villa, saluted smartly. Old Crone inclined her head.

'See that Manuel gets a nice drink, Sergeant, and a place in the shade.'

The next ones to arrive were Maurizio Capponcini and Jason Reed. Maurizio had picked up Jason at the Consulate. They had had a bite of lunch together at the Yacht Club. Jason had been irritable and had drunk two large vodkas. He had moaned about having lost Kate Harvey, just when they were really getting somewhere, and had cursed Bruce Winthrop the fourth roundly.

'We have a new arrival at the Club, Jason. She might well be your consolation prize. She's also on her own, and looks slightly forlorn. Why don't you come to our weekly cocktail bash, and I'll introduce you.

'Oh, I don't know, Maurizio.' Jason took off his glasses and polished them vigorously. That was always a sure sign that he was agitated. 'Let's face it, I'm just not lucky in love.'

'We'll see about that, but jump out now and I'll go and park the car.'

They gathered in Maria Beatriz's study, where the pencils and notepads had been distributed on the polished conference table. They were the inner circle and met regularly with Maria Beatriz and Sorciero to discuss the problems of Tequila. The preceding days had been so fraught with tension and Happenings, that the President and Sorciero hadn't yet been informed of the new developments concerning the Mangorenian exploration conducted by Sorciero. Now was the time to tell them about the properties of the Mangorenian spring, and how best to exploit them.

'We know that you have some interesting things to tell us, Old Crone,' Maria Beatriz started the meeting. 'You wanted the Bonheurs to join us later on, and they're waiting to be called.'

'Old Crone, what's this all about,' Sorciero asked. 'I didn't find any oil in Mangorenia; Etienne paid me the money, which I took, although it made me feel a bit guilty. So what else is there to discuss?'

'It's not the oil, Sorciero. It's the water,' Old Crone answered and looked at Jason. 'The consul made some inquiries on my behalf. He sent the water off to be analysed and we have had the results.'

'I don't understand; what's the big deal?' Sorciero pulled a face. 'It's not all that good; it's got a strange taste, but I suppose one gets used to it, when one lives there, don't you agree, Sweetness?'

Maria Beatriz wrinkled her brow. 'At least it didn't give me diarrhoea, but it's not particularly tasty. It's good enough for local consumption, maybe, but gimme a Perrier any time.'

'That's where you're wrong Madame President. It is far too good for local consumption. In future every drop of it should be saved for export.'

Maria Beatriz and Sorciero looked mystified. 'And that's all we have to tell the Bonheurs? I bet they've thought of export and decided against it.'

'Maybe, but they don't know what we know.' Jason entered the conversation.

'So tell us what you know, Mun,' Sorciero sounded impatient.

'The water contains the strongest aphrodisiac known to man. It will put all pharmaceuticals to shame. It was Old Crone who recognized the tiny flower growing all around the spring. They were used in an ancient Chinese remedy for impotence.'

Sorciero stared at Jason, and then started to laugh. It was an infectious laugh and soon they were all chuckling.

'Oh Mun, that explains it all. The permanent hard-on, the attraction to every female; all this fucking and all the babies born since the earthquake.' Sorciero slapped the table with glee. 'Even the chambermaid, at the Villa who was a real dog, seemed attractive...

'You didn't fuck her, Sam, did you?'

'Naw, Sweetness, but I wanted to.'

'That's the whole point,' Jason said excitedly. 'Not only is the water an aphrodisiac, but it looks like it's a fertility agent as well.'

'Bingo, I'd say. Almost as good as oil. Less messy and more fun. It might explain this mad passion you have developed for a certain lady recently arrived on the island, Jason,' Maurizio teased.

'So what do we do next?' Maria Beatriz asked.

'First, the water has to be fully analyzed for its properties, and evaluated by an appropriate scientific body. It must be given the seal of approval that it is fit for human consumption. Then the flow must be measured, to determine how many litres per minute it produces.' Jason said.

'It pisses like crazy,' Sorciero said enthusiastically. 'There's a pond full of it.'

'How can we get a cut of the action,' Maria Beatriz wanted to know. 'After all, we discovered it, didn't we?'

'Strictly speaking, it was Old Crone who came up with the goods,' Jason put in.

'True,' Maurizio said. 'We could put it all on a business-like basis, form a company, in which we will all be shareholders. Say on a fifty fifty basis with Mangorenia. They will need the capital which we can supply for building a proper bottling plant, design attractive bottles and labels, and deal with distribution.'

'No one will give a fuck for the attractive packaging when they will feel the result,' Sorciero crowed.

'We should have a good Hotel and Spa, on the best beach. A Club Royale on a larger scale. There will be an enormous influx of tourism, and there will have to be some additional accommodation. People will come from all over the world to drink the waters, like at

Evian, or Montecatini. How does that sound to you all?'

'Sounds cool to me, Mun,' Sorciero was running up and down the room as excited as a schoolboy.

'However, we want to keep it elegant and expensive, don't we,' Old Crone suggested.

'Natch; we don't want it to be like Coney Island,' Jason added his view and grinned.

'The exports will bring the bulk of the income. Marketing, marketing marketing, guys.'

'What is important is that the locals are weaned from the water, and put onto ordinary drinking water. The government will make enough money to give out free cokes and lemonades,' Mei-Ling made her contribution to the discussion.

'Good point. The water must be used exclusively for the Spa and for export.'

'They could make a sparkling drink with their Mangoreen juice. How about that?'

'I see that the possibilities are endless,' Jason pronounced. Maurizio looked round the table. 'Are we then agreed as to how to proceed? If we are, we should put it to Etienne Bonheur now.'

They all nodded their approval and raised their hands.

'But suppose that Etienne Bonheur tries to cut us out?'

'We won't mention the word aphrodisiac until we have signed the contract. Besides he needs our money and our expertise,' Maurizio said decisively.

'So what are we going to call this extraordinary liquid?' Mei-Ling asked.

'Cool Water, Mun, that's what; fucking Cool Water!' Sorciero crowed and danced around the room clapping his hands.

'And so say all of us,' the others around the table chanted.